MACROENGINEERING
An Environmental Restoration
Management Process

Also Available from CRC Press

Bioremediation of Recalcitrant Compounds edited by **Jeffrey Talley**

Chromium(VI) Handbook edited by **Jacques Guertin et al.**

Coastal Lagoons: Ecosystem Processes and Modeling for Sustainable Use and Development edited by **Ethem Gönenç and John P. Wolflin**

Ecological Risk Assessment for Contaminated Sites by **Glenn W. Suter II et al.**

Forests at the Wildland-Urban Interface: Conservation and Management edited by **Susan W. Vince et al.**

In Situ Remediation Engineering by **Suthan S. Suthersan and Fred Payne**

Practical Handbook of Environmental Site Characterization and Ground-Water Monitoring, Second Edition edited by **David M. Nielsen**

Restoration and Management of Lakes and Reservoirs, Third Edition by **G. Dennis Cooke et al.**

The Economics of Groundwater Remediation and Protection by **Paul E. Hardisty and Ece Özdemiroğlu**

MACROENGINEERING
An Environmental Restoration
Management Process

John Darabaris

CRC Press
Taylor & Francis Group
Boca Raton London New York

CRC Press is an imprint of the

CRC Press
Taylor & Francis Group
6000 Broken Sound Parkway NW, Suite 300
Boca Raton, FL 33487-2742

© 2006 by Taylor & Francis Group, LLC
CRC Press is an imprint of Taylor & Francis Group, an Informa business

First issued in paperback 2019

No claim to original U.S. Government works

ISBN-13: 978-0-367-45368-8 (pbk)
ISBN-13: 978-0-8493-9202-3 (hbk)

Visit the Taylor & Francis Web site at
http://www.taylorandfrancis.com

and the CRC Press Web site at
http://www.crcpress.com

Library of Congress Cataloging-in-Publication Data

Catalog record is available from the Library of Congress

The Author

Formerly a division vice president with Kearney/Centaur, John Darabaris is an experienced program manager on complex, sophisticated DOE, DOD, EPA, and industry environmental projects. Possessing both a professional engineer's (PE) license and a nonpracticing Certified Public Accountant (CPA) certificate, he marries both engineering and cost perspectives to the impacts of regulatory strategy alternatives.

With a background that combines graduate degrees in Geologic Engineering (MS, University of Missouri at Rolla) and Finance (MBA, Columbia University), Mr. Darabaris provides unique insights on the breadth of technical, regulatory, and management issues that program and project managers face in today's complex environmental corrective action management world.

In recognition of his achievements, Mr. Darabaris has been awarded an honorary Professional Development degree from the University of Missouri at Rolla and a commendation from the U.S. Army Corp of Engineer's Omaha, NE, office.

Preface

The purpose of this text is to provide the reader with some insight into the wide scope of subject matter that a project or program manager typically will face on a complex, large-scale environmental restoration project.

It has been my experience that few environmental professionals are fully prepared for the range of subject matter and issues that they will face as they progress through their careers into the ranks of project and program management. My aim is to provide junior and middle ranks, as well graduate programs, with a manual that, in a fashion, raises all the issues that a project or program manager will face.

Recognize that each of the subjects addressed, if dealt with at its proper depth, is a text unto itself. My goal is to provide a starting point and to also stress the interconnection between the key elements (e.g., remediation design and regulatory strategy need to be tied together, etc.).

Also, please realize that when I present specific examples (e.g., models, regulatory options, etc.), many of the details will be out of date before the ink dries. Models are continually being revised and improved, regulations are continually being redefined, site characterization techniques and mobile laboratory equipment are continually being improved. The point is not necessarily the specifics but the identification of the need for consideration of these issues, how they play out in the wider view of things and a stronger understanding of the integrative nature of all these separate items.

In addition, although I do provide some discussion into specialty areas (for example, unexploded ordnance), I have written the text to be universal in its applicability. In that sense, my hope is that it provides some useful management reference points for DOE, DOD, EPA, and industry led environmental restoration projects and programs. I also hope that it is written clear enough that it also provides insight that might be useful to less technical, tangential investment, insurance, and stakeholder communities

environmental remediation activities within the U.S. from site characterization, planning to implementation. With that background in mind, it is my hope that the text provides a broad perspective. It is not written from a "regulator" perspective or a "regulated" perspective, but from the simple perspective of "getting the job done" in an efficient, cost-effective, well-organized, and defensible fashion.

Contents

chapter 1

Macroengineering as an integrated environmental restoration management process

1.1 Introduction

Environmental restoration is celebrating its 30th anniversary worldwide, in recognition of the enactment of the Resource Conservation and Recovery Act (RCRA) in the U.S. in 1976. The nation is restless over the manner in which environmental cleanup is being conducted; criticism is coming from capitols, legislatures, and Congress. The *status quo* is under attack for a variety of reasons and rationales. The entire hazardous waste management and cleanup process, the finest to be found and internationally considered the standard of excellence, is being held up for scrutiny. The hue and cry is for more efficient cleanup approaches, particularly from a large-scale perspective, and for better control over unique environmental restoration challenges (unexploded ordnance [UXO], radioactive waste management, and cleanup).

From this debate, a window of opportunity is opening in the field of environmental restoration. At Congress' urging, EPA is evaluating accepting a more "risk-based environmental restoration approach and encouraging more flexible municipal–industrial cooperative brownfield restoration arrangements to remediate contiguous blighted urban areas on a timely, cost-effective, and realistic basis." As a result, the emphasis is changing from a legal-dominated, fault-finding exercise, to a paradigm of "get it done" in an expeditious manner exercise. The latter emphasis offers industry the opportunity to proactively reconstruct their environmental restoration programs

2   *Macroengineering: An environmental restoration management process*

mechanisms that encourage quicker RCRA-driven corrective action. In particular, the corrective action management unit (CAMU) rule offers industry the opportunity to undertake major RCRA-required cleanup actions without necessarily triggering land disposal restrictions.

Furthermore, the prior financial advantages of delaying cleanup through legal strategies may no longer hold true in the current era of lower interest rates and greater potential regulatory flexibility. In point of fact, there may not be a better time for resolving long-standing cleanup issues.

However, apart from the regulatory-driven and financially driven reasons for acting, the record is now clear that environmental restoration costs and natural resource damage (NRD) costs will explode unless careful, up-front strategic planning of an integrated nature occurs, followed by timely self-examination and ongoing environmental restoration management control.

Proactive planning is not only possible but economically attractive through a macroengineering approach.

Macroengineering represents the assumption of management control over environmental site restoration by developing an integrated plan for site and waste characterization and risk assessment based on planned future use. Issues are identified, flagged, solved, and negotiated on a priority basis, in frequent, constant, direct contact with regulatory personnel, so that perturbations from personnel turnover or regulatory drift are minimized.

As shown in Figure 1.1, uncertainties drive the need for an integrated environmental restoration approach that maps out a realistic strategy and defines an achievable end product.

Uncertainties are project impacts nominally lying outside the control of project management. Uncertainties relate to unresolved issues or undeclared agenda or responses by parties to the remediation process. Macroengineering seeks to identify, early-envelope, and convert uncertainties to known factors that can be included in the overall management plan.

Besides the technical, cost, and schedule uncertainties identified in Figure 1.1, regulatory uncertainties also play a significant role in driving program uncertainties. The Superfund legislation of the 1980s provided the

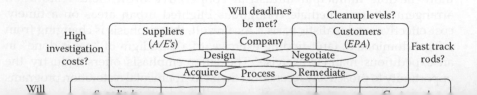

impetus for promulgation of Natural Resource Damage Assessment (NRDA) regulations. In the past several years, the NRDA rules have undergone several major revisions and been subject to legal rulings (e.g., *Ohio v. U.S. DOI*), the net effect of which could potentially increase the dollar value of natural resource injury claims, if applicable restoration does not occur. The key factors driving this escalation are:

1. Expansion of what constitutes natural resources subject to damages.
2. Expansion of liabilities from "the lesser of restoration or replacement costs; or diminution of use values as the measure of damages ... " [43 CFR 11.35(b)(2)], to restoration or replacement costs plus the NRDs that occurred earlier and which will occur in the future.
3. Expansion of the value of damages to include nonuse values. Some measure of relief has been provided to potential responsible parties (PRPs) if they can prove that the restoration is unfeasible or the costs are "grossly disproportionate" compared to damages, and a spirit of action is presented.

The objective of the macroengineering environmental restoration management process presented herein is to increase the overall effectiveness by which organizational resources, committed to environmental restoration, are utilized. In essence, macroengineering is a management program to effectively integrate regulatory, technical, and management issues to provide well-rounded, cost-effective environmental restoration solutions for large-scale restoration projects.

The focus of macroengineering is not limited to overall environmental management goal setting, but includes establishing detailed technical planning, regulatory documentation, and cost estimation protocols to ensure the desired results are achieved. Although undertaken from a senior management perspective, macroengineering also encompasses detailed preparation of critical environmental regulatory documents (records of decision, remedial investigation and feasibility studies, environmental permits, etc.) and technical information (monitoring data, sampling plans, risk assessment studies, etc.) from the standpoint of their strategic value, given cost, schedule, and regulatory objectives.

Macroengineering takes a system-based, "big-picture" environmental restoration management approach to its review. Under a macroengineering process, select activities are not treated as individual units, but as a part of a total view to environmental restoration problem identification and resolution. As a result, the process generates a greater understanding of potential resource requirements and the impact of technical/regulatory hurdles ("showstoppers") on meeting remediation goals.

4   *Macroengineering: An environmental restoration management process*

- Establishment of an environmental baseline engineering document
- Development and review of policies, guidelines, and procedures relevant to establishing technical approaches and controlling technical quality
- Development and review of cost- and schedule-estimating processes
- Independent cost and schedule review of a statistical sample of projects across the site's environmental restoration site universe
- Establishing the approach and review process for a statistical sample of monitoring data to ensure compliance with data quality objectives and cost-effective regulatory strategy
- Evaluation of the site remediation contract options for their ability to control contractor activities from a technical, cost, and schedule standpoint
- Evaluation of the control processes for activities funded by indirect charges under site remediation contracts
- Evaluation of the technical and regulatory decision-making process and documents prepared or to be prepared
- Identification and assessment of regulatory/technical impacts on cost and schedule via value engineering and cost benefit studies
- Identification of contingency management and enhanced cost control opportunities

The process can be used to address the adequacy by which the site's environmental restoration program is dealing with the issue, both corporate-wide (in the case of multiple sites) and at each individual site. Obviously, there is a need to reflect on the different programmatic needs for a given site.

A central question to ask is whether the company is better served in considering the environmental restoration activity as a program versus as a project. Inherent within the title "program" is a greater emphasis on development of internal resources for managing the mission via staff development and equipment acquisition. Perhaps the main factor in determining this is if the company (agency) owns or is involved in more than one site and there is, or can be effected, an agreement with the regulatory agency to allow a string of separate cleanups over one or two decades. In such case, a corporate level agreement may not only save money but could be used effectively to tighten up the restoration effort, making it more responsive to corporate goals. However, environmental restoration, in most cases, is a unique mission outside the mainstream scope of most corporate activities. Thus, it may be better for companies to consider the environmental restoration mission as a project management exercise in which technical resources are, by and large, contractor-supplied and the company's environmental restoration is

Chapter 1:   Macroengineering as an integrated environmental restoration   5

	Program management emphasis	Project management emphasis	
	Industry demands a heavy investment in environmental health and safety	Industry does not typically demand a heavy investment in environmental health and safety	
Unique problem: Value-added support required	**Exceed standards** Develop best capability internally	**Develop access to** Best capability within a cost/benefit	• Significant mission scope (size)
Standard problems: Basic support required	Meet standards	**Develop access to** Capability that ensures compliance	• Limited mission scope (size)
	Concern over proprietary issues	No concern over proprietary issues	

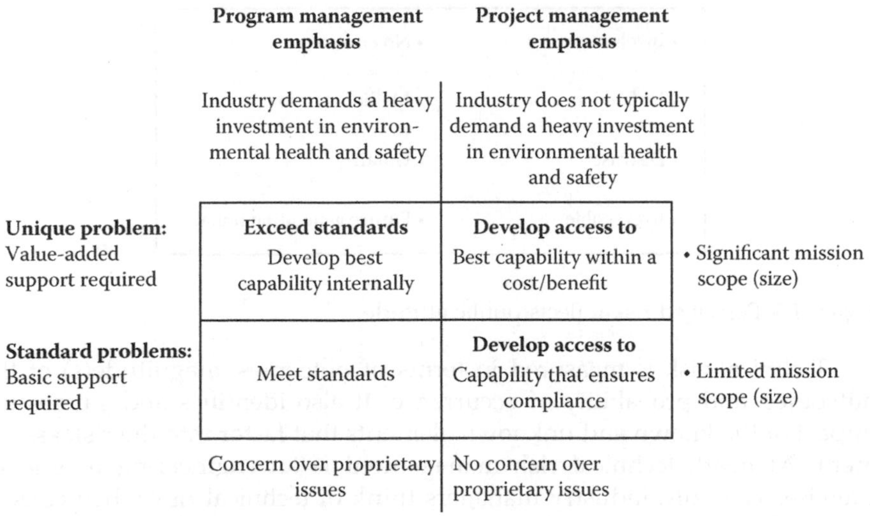

Figure 1.2 Program versus project management analysis.

In areas in which heavy emphasis is placed on utilizing outside subcontractor resources, a company's environmental restoration management philosophy should be structured so as to maximize the potential for sharing cost/schedule risk and management risk with subcontractors under well-designed incentive programs. This can be accomplished by addressing two issues: an independent NRD assessment element and a "managed risk" assessment that includes a public participation element and provides the company with independent feedback of key issues that define the ultimate success of its environmental restoration program at a given site.

The thrust of the NRD assessment activity is "How to avoid being a target of NRD"; or, if you cannot avoid becoming the target of a claim, at least do the best job you can to prepare and position yourself effectively. The assessment entails finding out (through knowledgeable third parties) whether any trustee agency has initiated an NRD review and (if so) what its review criteria and priorities may be, identifying others in the "same boat," as well as determining the basis for and scope of the claim. Chapter 8 discusses this issue in detail.

The second element is managed risk assessment.

From a management perspective, there are three types of environmental risks:

6   *Macroengineering: An environmental restoration management process*

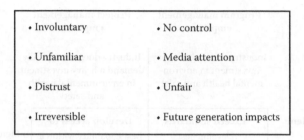

• Involuntary	• No control
• Unfamiliar	• Media attention
• Distrust	• Unfair
• Irreversible	• Future generation impacts

Figure 1.3 Perceived risk reflects public attitude.

Technical risk is measured in terms of outcomes, magnitude(s) of the outcomes, and probability of occurrence. It also identifies and gauges the impact of the known and unknown elements that factor into the risk assessment. At heart, technical risk management is an engineering or science exercise. Typically, industry managers think of technical risk when considering risk management.

Perceived risk reflects the public attitude. It is critical to understand that the public's ranking of perceived risk is not highly correlated with actual technical risk. Figure 1.3 presents a suite of issues and concerns that could fall within the perceived risk assessment. However, each site is unique, and its perceived risk will reflect its unique conditions. Perceived risk is perhaps the most difficult of all risks to anticipate and to deal with efficiently. It can be manipulated by activist groups interested in controlling or defeating the site restoration project. Perceived risk can be skillfully modified and presented to arouse the sympathy of the media, public, politicians, and the regulatory community. The last target group generally is resistant to this type of pressure, however, and if the restoration approach is not well planned and presented, it may suffer defeat in the public forum.

Regulatory risks deal with compliance with current federal and state standards as well as compliance with corporate (and in the case of DOE, DOD, other regulatory governmental agency (e.g., EPA, OSHA, NRC)) policies. Regulatory risk analysis should be approached with a proactive attitude. This means assessing the potential for retroactive application of future standards (e.g., Superfund) and the latter's implication on the site. It also means development of future internal standards reflecting the public's risk perceptions, rather than actual risks.

Again, it should be recognized that future regulatory risks typically are politically driven; they do not necessarily reflect technical risk. However, regulatory risks are the ones that ultimately dictate cost to a far greater degree than technical risk and, as such, require as much emphasis, if not more

Chapter 1: Macroengineering as an integrated environmental restoration   7

process of developing an understanding of the business consequences of each risk component. Prioritizing risks is a process of combining the technical and regulatory risks with the perceived risks. Identifying risk management alternatives involves not only assessing the technical and regulatory options, but also the outreach/communication options available. In some cases, alternative strategies may require combination of technical, outreach, and other elements.

The alternatives should then be analyzed for their cost/benefit and also from the standpoint of the uncertain vs. certain composition of the cost/benefits. Implicit within this analysis is the crucial question of feasibility — from the technical, regulatory, and public acceptability points of view, all of which impact cost feasibility. Central to the analysis is the question, "How much are we willing to pay for the 'uncertain benefits' of a strategy?" Also, how much uncertainty are we willing to live with, relative to the cost, for specific options?

The question must be asked, "How is this risk of failure in the remediation concept to be measured?" Central to the measurement of the potential failure of a remediation program is the technology risk assessment (TRA). The latter can be defined in a systems approach to identify and evaluate the risks associated with a given remediation or waste management technology. A TRA must consider risks associated with technology failure, indirect consequences, and primary and secondary risks of accidents and malfunctions. TRA does not duplicate or replace human health or ecological risk assessments, but supplements them with the boundary condition of realistic technology expectations.

Thus, the selection and implementation of a strategy, particularly for complex risks inherent at sites that lend themselves to the macroengineering approach, will require actions at multiple levels.

In short, environmental restoration risk management requires integrating risk management analysis with the key business management processes of the organization. It also involves actively managing technical, perceived, and compliance risks. Environmental risk management is part of the company's or site's job and must be integrated into the overall environmental restoration management processes.

chapter 2

Preparation of a preconceptual engineering baseline study

The preconceptual engineering baseline study will be the core of the macroengineering process and will involve a detailed development and ongoing review of the environmental restoration program — both from an overall management control standpoint and a detailed technical and regulatory standpoint. As shown in Figure 2.1, the preconceptual engineering baseline study should be developed, based upon several in-house models that will define both the site's technical and regulatory status and the overall programmatic effort required.

These models are a *site technology assessment model*, a *site organizational model*, and a *site regulatory model*. They will be the basis for evaluating the site's environmental restoration management posture and establishing the overall effectiveness of the organization's environmental restoration program, from a site control perspective.

The site technology assessment model and the site organizational model should draw upon proven systems engineering methodologies such as detailed definition of system processes and procedures, mission analysis, functional analysis, development of functional flow diagrams, requirement definitions, criteria selection, analytical tool development and selection, system engineering decisional analysis methods, and buyer systems engineering facilitation methods.

The site technology assessment model provides the analytical approach needed to establish each site's baseline for action. As shown in Figure 2.2, this process evaluates the given site's readiness to move forward in its environmental restoration program from a technical system standpoint.

Any or all of the elements identified in Figure 2.2 suggest a reason for a macroengineering approach. The "macro" approach addresses all aspects of systems engineering support with the objective of "getting our arms

Baseline engineering study

| Site technology assessment model | Site organizational assessment model | Site regulatory model |

⇑   ⇑   ⇑

| Technology survey element | Natural resource damage assessment | Regulatory and public perception scan |

Figure 2.1 Interrelationship of elements and models within the macroengineering framework.

Some of the technologies to consider include large-scale excavation and material-handling equipment; remote sensing, testing, cleaning, cutting, and excavating; advanced treatment methods for contaminated soils and ground-water, including *in situ* stabilization technologies; and innovative design for waste repositories and disposal sites. The goal of macroengineering is to clearly identify in a preliminary fashion what it would take, in terms of manpower, equipment, cost, productivity, and regulatory support, to meet the milestones and goals of a given large-scale environmental restoration mission.

This step of the process will identify critical gaps and inconsistencies in the site environmental restoration management practices. Furthermore, this

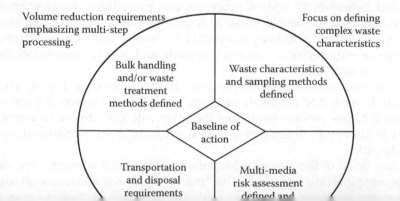

Chapter 2:   Preparation of a preconceptual engineering baseline study   11

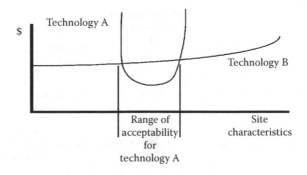

Figure 2.3 Tradeoff between technologies, site characteristics, and cost.

step provides a baseline for evaluating the potential impact of other technology development options that may be on the horizon.

This step will also attempt to identify the potential site-characteristic
sensitivities that may eventually impact cost and schedule. As shown in
Figure 2.3, there are potential significant cost tradeoffs between technologies,
site characterization, and site characteristics. Whereas Technology A may
offer cost advantages at a given site characterization range, Technology B
may offer cost advantage vis-à-vis the less expensive site characterization
requirement necessary for its implementation, and given its lesser sensitivity
to given site characteristics, etc.

The site organization assessment model is an analytical model to develop
the site's organizational systems in a manner that effectively utilizes the
human and technology resources available, as well as adequately addresses
the contingent technical, regulatory, and management issues. In addition,
this model will attempt to address the degree of flexibility in the site's
organizational approach for dealing with possible site-characteristic sensitivities identified in the site technology assessment model. As shown in
Figure 2.4, uncertainties drive the need for a macroengineering approach. In
essence, macroengineering is a baseline to map out an environmental restoration strategy and define an end product that pierces the veil of uncertainty.

To that end, regulatory issues have profound effects on the organizational realities of an environmental restoration program, directly influencing
technical and schedule, and indirectly, but decisively, influencing cost. However,
regulatory-driven organizational approaches are not necessarily optimal
from the standpoint of long-term goals and performance objectives. Furthermore, from the company's standpoint, cost is the key resource limitation and
has a direct influence on the technical options, schedule, and, ultimately, the

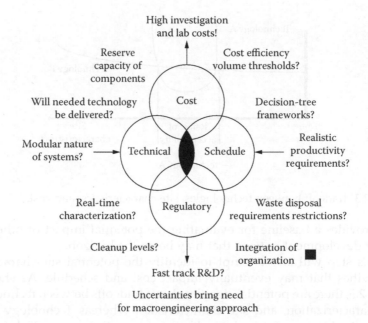

Figure 2.4 Site organization assessment model framework.

program efforts. Regulatory climate assesses the public perception relative
to a specific site's criticality and a company's overall role as an ombudsman
for the public environmental trust, as well as the regulatory infrastructure.
The latter may be well established. In some respects that can be beneficial,
because regulatory requirements may then be well defined in contrast to an
undefined regulatory environment that leaves site cleanup, from a political

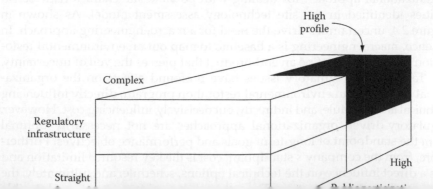

Chapter 2: Preparation of a preconceptual engineering baseline study 13

standpoint, undefined. On the other hand, we are all familiar with the debilitating effects of needless, complex, and confusing regulatory requirements. We add a third element — the language and documentation stiffness factor. This reflects the complexity of the documentation requirements that may be required — the complexity of language translation requirements in the case of international environmental restoration programs, and the challenge of translating from "technospeak" into plain English in domestic cases.

Thus, macroengineering is a system-engineering-based alternative that emphasizes the use of large-scale equipment to clean up large areas of contamination with emphasis on minimizing the extensive site characterization that is currently tying up numerous sites from timely cleanup. Under macroengineering, prerestoration site characterization will focus on screening rather than definitively characterizing. To explain it further, preremediation site characterization will be limited to identification of waste characteristics that can be used to validate screening criteria, the latter implemented under an observational approach and/or a mobile laboratory data-generated basis. The feasibility of the macroengineering approach to a given site is based on determining the availability of best acceptable technology (BAT). BAT is applied to the general waste characteristics of the site, whether the circumstance lends itself to large-scale modular operations that can reduce cost via economics of scale and also reduce potential risks from uncoordinated environmental restoration efforts. As such, the macroengineering approach should also include the identification of critical technology and engineering developments that may enhance the feasibility of this approach.

In the U.S., for example, the current technical approach to remediation at most sites is driven by regulatory requirements defined by the EPA and state environmental agencies. Site-specific regulatory agreements based on current regulatory authorities typically force the independent cleanup of individual units in a piecemeal fashion that has a microengineering orientation. Remediation of the individual waste sites is directed by either the Resource Conversation and Recovery Act (RCRA) or the Comprehensive Environmental Response, Compensation, and Liability Act (CERCLA) authorities.

Based on the site characterization under the CERCLA authority, engineering feasibility studies are prepared for each unit. This ultimately leads to a record of decision (ROD) identifying the selection criteria and the remedy chosen and approved by the regional USEPA administrator. In cases in which time has lagged and new information is gathered or new technologies have become economically viable, the ROD may be recognized or legally challenged as obsolescent, and a revision may be made as an explanation of

14   *Macroengineering: An environmental restoration management process*

by (1) removal and treatment of wastes, (2) demonstrations of clean closure, or (3) closure by application of standard cover followed by long-term postclosure maintenance and monitoring, as laid out within a correction measure study.

Thus, investigation of all units under either RCRA or CERCLA typically requires extensive and duplicative time, effort, and costs prior to the implementation of actual cleanup. Furthermore, differences in implementation and conclusions (such as cleanup levels) may occur because of the different regulatory authorities involved (e.g., RCRA and CERCLA). This can be especially confusing (and inefficient) for those sites that have both an RCRA and CERCLA component, which is typical for large complex sites where macroengineering is most applicable.

Implementation of the current microengineering approach to cleanup may present several other negative aspects when applied to a large site. Remedial investigations that focus on small segments will delay the emergence of the total picture of cleanup requirements at large sites. This delay makes it difficult to consider large-scale, integrated remedial approaches, such as the identification and acquisition of common treatment facilities or processes to be applied to all (or most) waste sites. Delays may also result in an increased public demand for more visible, robust cleanup. The regulators may respond by requesting interim cleanup actions at individual sites or operable units that are, at best, stopgap and may lack in integration with overall site restoration goals. Also, by focusing on smaller operable sites without due consideration of nearby units of similar nature and the type of waste, the selected remediation technologies may vary significantly from unit to unit. Variation will preclude cost savings based on economies of scale and will result in the use of different types of unrelated, nonintegrated, and nonreusable equipment. Further cost savings will be lost because a learning curve must be established for each new technology.

Current experience has clearly shown that one must manage not only the present regulatory circumstances but also the future, via systems-engineering-based management approaches that clearly articulate and integrate regulatory strategies, cost requirements, and scheduling challenges. The goal is to clearly identify "show stoppers" and identify clearly, in advance, the need for prior negotiation and renegotiation of interagency regulatory agreements to achieve overall programmatic goals.

Macroengineering approach is a preconceptual engineering baseline study that deals with the development of concepts and technologies for cleanup by aggregate area and sitewide groundwater remediation that emphasizes the use of large-scale mining and industrial engineering approaches. The goal of a macroengineering approach is to significantly

Chapter 2: *Preparation of a preconceptual engineering baseline study* 15

Table 2.1 Comparison of Key Elements of Technical Approaches

Macroengineering	Microengineering
Cleanup by aggregate area	Cleanup by operable unit
Use large-scale equipment defined up front	Use small-scale equipment defined site-by-site over time
Use common facilities defined up front	Facilities defined site-by-site over time
Minimizes up-front investigation	Maximizes up-front investigation
Waste disposal defined	Waste disposal undefined
Primary product: contaminated land area cleaned per dollar	Primary product: investigation, documentation and costly data evaluation and piecemeal cleanup

using large-scale equipment, and minimizing preremediation characterization by emphasizing the observational method, supported by real-time mobile laboratory analysis. Typically, each large-scale area should comprise multiple waste units that were once dedicated to a common mission and thus are likely to contain similar waste types. Table 2.1 compares the elements of macroengineering and microengineering.

Common facilities may include waste treatment facilities, safety and health control, on-site transportation and waste handling operations, on-site laboratory services, and on-site waste packaging.

In essence, macroengineering is a system's engineering-based management approach to sitewide environmental restoration. It emphasizes the marrying of regulatory, technical, and management experience to provide a well-rounded, total view to environmental problem identification and resolution. Critical to the macroengineering approach is developing a thorough understanding of the problems associated with the given environmental restoration mission and quickly incorporating the various initiatives, that may be in progress, into the problem resolution process. The macroengineering approach emphasizes obtaining a "big-picture" understanding of the potential resource requirements, technical/regulatory hurdles, and remediation goals to provide focus and coordination in the development of the more detailed studies to follow.

The typical macroengineering baseline study should present a conceptual design, environmental impacts, modeling, cost estimates, emergency response analysis, and innovative technology development opportunities are provided. Within the study, a proposed protocol for establishing cleanup standards. Also, the study should present mass balance process flow dia-

16   *Macroengineering: An environmental restoration management process*

The initial preconceptual engineering baseline study provides the basis for negotiation of a site's regulatory agreement. Evaluation and comparison of the various technical approaches under consideration lead the macroengineering effort toward establishing a common basis, along with formulation of a sitewide strategy that identifies target objectives for future land/groundwater. It is essential that the strategy balance the desire for restored lands with the realities of waste disposal. For instance, disposal of waste at sites other than the generating site may not be feasible from a public policy standpoint. In such cases, areas within the generating site will need to be reviewed as candidates for an on-site engineered repository. However, in doing so, a large area of a site may be restored to public use in a cost-effective and timely manner. This beneficial tradeoff should be assessed and reflected in public policy discussion.

As regulatory and technical negotiation and decision progress, four major follow-up actions to the original macroengineering study should be prepared:

- Project plan for implementing a test phase for macroengineering activities
- Cleanup objectives
- Regulatory strategy for macroengineering implementation
- Detailed cost estimates for macroengineering implementation

The project plan document focuses on the engineering/design, permitting, regulatory, and interface activities required to achieve macroengineering milestones.

The project plan should address the following items:

- The initial list of documents known to be required. This includes identifying and defining the planned contents of the documents, interfaces, decision points, and activities (e.g., data gathering, testing, modeling, etc.) required for document completion.
- The relationships between functional design criteria, conceptual design, and definitive design.
- The scope of each of the two design steps given in the preceding text (e.g., define the expected design inputs and outputs, review cycles, and hold points).
- The interfaces between the design functions as currently identified (e.g., excavation/extraction systems, containment system, material-handling and analytical systems, and disposal facility).
- The scopes and responsibilities of the permitting NEPA documents

Chapter 2: Preparation of a preconceptual engineering baseline study   17

- Organizational responsibilities for the various activities.
- Strategies for gathering needed data.
- A detailed, integrated project schedule.
- Project cost estimates.
- Quality assurance requirements.
- Document control.

The primary objective of this activity is to produce a project plan that will serve as a detail reference document for project participants and other interested parties. The project plan is designed as a management tool that provides guidance for the design and review process, interface control, document control, assignment of safety/impact levels, establishment of quality assurance and safety requirements, and cost and schedule estimates.

Sitewide cleanup cost estimates using the microengineering approach have typically proved to be costly and difficult to budget with any real level of accuracy. In addition, these efforts have typically proved to be inefficient (timewise) in achieving the required cleanup within publicly acceptable time frames. The high estimated costs (and lack of budget control) and lengthy schedules warrant consideration of alternative approaches that (1) achieve a high measure of cleanup at a significantly lower cost and (2) provide for cleanup within a schedule that meets the public's acceptable time frame while maintaining an acceptable level of worker and public safety.

This text examines the potential for, and impacts of, using a larger-scale remediation approach, defined as the *macroengineering approach*. In macroengineering, a site is divided into large operational (aggregate) areas for remedial purposes. Each aggregate area comprises a number of smaller, less complex, operable units that would typically be the study focus on an individual basis under the current microengineering approach. Throughout the history of the site, each operational area will have had specific missions, and similar contaminants may thus be found throughout the individual operable units of an aggregate site area. Therefore, an economy of scale may be achieved by applying macroengineering techniques at the aggregate area to implement cleanup.

Typical strategies to consider to facilitate the evaluation are outlined in the following:

- Where clean closure is economically feasible, there will be a bias for source areas sites to be excavated or otherwise recovered and the contents removed, consistent with optional potential future land use, such as industrial or agricultural. The exhumed wastes would

18   *Macroengineering: An environmental restoration management process*

administratively and physically in perpetuity, and alternate water
supplies will be provided for use in these areas.

- Source area sites that cannot be cost effectively clean-closed and/or
  are potential candidates for on-site TSDF area will be assumed to be
  isolated, stabilized, and disposed of *in situ*. Engineered barriers
  would be applied for effective *in situ* disposal of existing sites in the
  area and disposal of wastes transported to the area from other source
  areas within the aggregate site area.
- Any large, relatively uncontaminated areas of the site between source
  area sites will be assumed to be restored, remediated, or enhanced
  so as to remove physical hazards and isolated structures, and reno-
  vated for potential constructive uses.

Because macroengineering cleanup is approached as a large-scale indus-
trial/construction project, appropriate industrial/construction work efficiency standards can be applied with modifications to reflect the unique
health and safety concerns at a hazardous waste site. Care should be taken
particularly when dealing with labor productivity at federal facilities. The
latter may be a fraction of what comparable industry environmental remediation efforts maintain. Therefore, labor rules at macroengineering-type
federal facility sites may need to be evaluated and renegotiated to more
closely parallel overall industrial standards.

A separate assessment of emergency response situations should evaluate
the differential in overall public and worker risk using the macroengineering
remediation approach vs. the current, unit-by-unit, microengineering
approach. Typically, worker and environmental safety in these operations
will not present any unique problems, but will require the adaptation of
established monitoring, personnel protection, and decontamination procedures on a larger scale than typical remediation sites. In addition, the grow-
ing number of health and safety regulations relating to the handling of
hazardous chemical (including unexploded ordnance) and radioactive
wastes will no doubt impose, in the future, compliance requirements more
complex than those currently existing and, as such, may be more easily met
on a larger scale.

Macroengineering follows the general approach to environmental restoration worker safety. By that, site work zones must be established. Workers
are provided with a localized contained work environment where protective
equipment is not required, and supplemental support systems are provided
in the event that unplanned events and emerging conditions may require
evacuation. Work in high-contamination and contained areas should be performed to the maximum extent by using remotely operated equipment. The

Chapter 2: Preparation of a preconceptual engineering baseline study   **19**

The safety officer (industrial hygienist/health physicist) performs a hazard evaluation for each site and, based on this evaluation, sets up specific procedures, specialized equipment, and controls to be implemented to ensure protection of the workers, the public, and the environment. The defining nature of the macroengineering organizations lends itself to taking advantage of the learning curve and common facility advantages relative to the safety organization and facility developments.

An element critical to the safety program of the macroengineering approach is the development of a large-scale mobile site containment structure or equivalent alert suppression systems. The objective of the system design is to prevent the spread of airborne contamination to the environment during excavation. Where lesser dust suppression systems are used in lieu of containment structures, they should be accompanied by stricter real-time monitoring capabilities at the point of operation, as well as the availability of "hot" spot suppression systems such as mobile gunite equipment.

The baseline should also provide for development of a regulatory strategy document. The latter is an evaluation of all applicable environmental regulations that could bear on the implementation of macroengineering. The regulatory strategy document identifies and evaluates all relevant and applicable regulations, acts, and laws in relation to the implementation of macroengineering, including Department of Energy orders and Department of Defense orders, when operating within these federal agencies' environments. The regulatory strategy document deliverables include:

- A matrix that lists each act; summarizes the major provisions bearing on macroengineering; identifies provisions compatible and provisions incompatible with macroengineering implementation; identifies opportunity for exceptions, waivers, or negotiation; and identifies areas in which regulatory negotiation will be required and initial positions should be taken with regulators. Probability of success for incompatible provisions should also be provided along with case studies or precedents in which similar disposal/cleanup approaches have been applied and approved.
- A logic chart that develops a pathway through the regulatory maze that allows the successful implementation and emphasizes potential "show stoppers."
- An overall regulatory strategy that describes conclusions, issues to be addressed, recommendations, case studies, evaluations, and analyses in a coherent fashion.

20   *Macroengineering: An environmental restoration management process*

Bibliography

1. Adams, M.R., A Macroengineering Approach to Hanford Cleanup with Land Use and Technology Development Implications, WHC-SD-EN-AP-037, Westinghouse Hanford Company, Richland, WA, 1990.
2. Adams, M.R., Statement of Work (Revision 6), Preparation of Conceptual Hanford Site Cleanup and Restoration Engineering Study, Westinghouse Hanford Company, Richland, WA, 1990.
3. WHC, Hanford Past Practice Site Cleanup and Restoration Conceptual Study, Integrated Study and Summary, WHC-EP-0456, Vol. 1 and 2, Draft, Washington Hanford Company, Richland, WA, 1991.
4. U.S. DOE, DOE Order 5480.11, Radiation Protection of Occupational Workers, Washington, D.C., 1988.
5. U.S. DOE, DOE Order 5820.2A, Radioactive Waste Management, Washington, D.C., 1988.
6. Department of Energy, Richland Office, Westinghouse Hanford Company, *Radiation Protection Manual* (WHC-CM-4-10), Hanford, WA, 1988.
7. Occupational Safety and Health Administration (OSHA), 29 CFR 1910, 1986, Title 29, Labor, Part 1910, *Occupational Safety and Health Guidance Manual for Hazardous Waste Site Activities*, Washington, D.C., 1985.
8. Gustafson, F.W. and Green, W.E., Large-Scale Surface Excavation of Contaminated Waste Sites on the Hanford Reservation, WHC-SD-EN-ES-007, Rev. 0, Westinghouse Hanford Company; Richland, WA, 1990.
9. NIOSH, *Occupational Safety and Health Guidance Manual for Hazardous Waste Site Activities*, U.S. Dept. of Health and Human Services, Center for Disease Control and Prevention, National Institute for Occupational Safety and Health, Washington, D.C., 1985.
10. U.S. DOE, DOE Order 5400.5, Radiation Protection of the Public and the Environment, Washington, D.C., 1990.
11. Hutchinson, I.P.G. and Ellison, R.D., Mine Waste Management, sponsored by the California Mining Association, 1992.
12. U.S. DOE; the American Society of Mechanical Engineers, United Engineering Center, New York. Mining Workshop for Nuclear Waste Cleanup, sponsored by the Mining and Excavation Research Institute, ASME, U.S. Bureau of Mines, Colorado Center of Environmental Management, held in Colorado Springs, CO, 1991.

chapter 3

Macroengineering technical approaches

3.1 Contamination soil and source area site environmental restoration approaches

Macroengineering focuses on three approaches to large-scale soil/source area site environmental restoration: (1) a volume reduction approach based on treatment technology, otherwise known as the *industrial approach*; (2) a bulk-handling approach based on expeditious excavation and removal, otherwise known as the *mining approach*; and (3) a *contained approach* that emphasizes dedicated large-scale hazardous waste management and policy solutions. These are briefly summarized in Table 3.1.

These three approaches are not to be considered mutually exclusive, and common elements for both may be applicable to a given restoration program.

Typically, the detailed macroengineering soil and source area environmental restoration studies provide the following information:

- Engineered system concepts and descriptions of the alternative that will achieve the required degree of cleanup for each option. The engineered concepts should address the seven engineered components that apply, including: (1) site containment during recovery; (2) site recovery, including cutting of oversized objects in excavation; (3) waste treatment for volume reduction; (4) transport (and packaging) to an on-site repository; (5) on-site disposal and waste form (as applicable); (6) *in situ* disposal of sites; and (7) treatment or isolation of groundwater as applicable to each case described.
- The restoration of the sites following cleanup. Sites must be returned to a condition that allows revegetation while minimizing water in-

22   *Macroengineering: An environmental restoration management process*

Table 3.1 Macroengineering Study Variables

	Study area	Technology orientation	Future-use objectives
Mining approach	Soils with light and surficial contamination over a wide expanse	Straightforward surface-mining-based approach with no reduction or treatment of waste; emphasis on high-volume, bulk-disposal handling technologies	Option 1: site restoration levels that will support general use Option 2: site restoration levels that will support industrial use
Contained approach	Significant amounts of buried waste and complex disposal unit storage	Straightforward disposal facility and barrier technologies	Optimal site remediation to support industrial use in some areas
Industrial approach	Highly variable waste-disposal units and significant decontamination and decommissioning requirements	Flexible environmental restoration emphasizing multistep environmental process engineering approach with increased waste characterization; emphasis on waste reduction and treatment, on low bulk, and on handling and disposal methodologies	Option 1: site restoration levels that will support general use Option 2: site restoration levels that will support industrial use

- A rough schedule for completion of each concept for each option.
- Identification of engineering development tasks and tests required to bring each concept to fruition. (Modifiable available technologies have only been considered for engineering development; where no technologies exist, technology development needs must be defined.)

- Apply remediation actions to entire blocks of sites or aggregate areas rather than to individual sites.
- Utilize high-throughput equipment to clean or treat large volumes of contaminated soil or waste rather than sizing equipment to clean or treat volumes representative of smaller, individual sites.
- Excavate, move, and treat large volumes of soil or waste removed from sites. Rely on high-throughput treatment, detection, and sorting techniques to treat, clean, and volume-reduce these materials for separation of the contaminated portion prior to disposal.
- Apply treatment processes that are standard technology, not temperature or pressure dependent, and that have low energy consumption with a minimum of secondary-waste generation.
- Are insensitive to (not dependent on) extensive prior site knowledge, lack of detailed site design, the type or concentration of contamination present, the physical state of deterioration of the site, the site volume or area, the physical nature of the soils and debris found in the site, and proximity to populated areas.
- Apply techniques that are insensitive to (not dependent on) extremely fine definition of contaminant plume locations or concentrations either in the soil or groundwater. This concept accounts for the reality that such detailed characterization may not be economically feasible at most large sites.
- Isolate (or treat) groundwater contaminant plumes on a wholesale basis rather than site by site. Consider potential redirection of the overall groundwater flow regime at the site by use of engineered structures or facilities.

Macroengineering emphasizes a decision tree approach to control field remediation activities. In this scenario, regulators play a role in establishing the decision tree framework, not in running the remediation. Specifically, to conform to the overall engineering approach, the engineering techniques should be selected in consideration of the following criteria to the extent practical:

- Require minimal, detailed prior site knowledge as to the nature and extent of contamination in any given site. The investigation techniques that are critical to implement the engineered systems should be fast, flexible, low cost, and an integral part of the engineered system.
- Emphasize mobile/transportable waste processing to deal effectively with scattered, dispersed geographic locations of waste units at large-scale restoration sites.

24   *Macroengineering: An environmental restoration management process*

premium on in-field, real-time analytical techniques providing rapid response and including geophysical techniques.

- Take advantage of the general tendency of contaminants to associate within a given component of the site media (e.g., a preference for accumulation within the fine fractions of a soil column).
- Return as much cleaned material back into the excavated site as possible.
- Restore the cleaned-up sites to a condition suitable for a variety of industrial or general uses.
- Minimize secondary-waste generation.
- Minimize equipment decontamination requirements.
- Minimize both capital and operating costs.
- Protect workers and the environment during cleanup operations and comply with all regulations and standards for worker protection.

Significant differences in remediation unit costs typically occur between systems that emphasize the industrial approach vs. systems that emphasize the mining approach. Although both systems may propose similar excavation technology, such as major mining excavation equipment and systems coupled with mobile containment systems, and both approaches would rely heavily on mobile laboratory support to provide timely analytical data to direct field cleanup operations without undue delay, there typically is a significant technology divergence in processing and packaging the excavated material. In the mining approach, no waste processing other than segregation and packaging transportation is utilized. The industrial approach, on the other hand, will incorporate waste treatment steps to reduce waste volumes, such as super compaction, soil washing, and low-temperature thermal desorption. In addition, given that the industrial approach is probably more applicable in cases in which there is a greater variety of waste types expected, the latter system will require more extensive segregation and a wider variety of special waste containers.

3.2   *Mining-oriented macroengineering approach*

In the mining-oriented macroengineering approach, the key elements are to excavate rapidly, containerize the wastes, and transport them in an environmentally safe manner, all at minimal unit cost. The emphasis is on simplicity, using currently available techniques when possible, such as those practiced in the mining industry and in the uranium mill tailings environmental restoration programs. Thus, the mining-oriented approach typically emphasizes bulk handling and would not utilize the more com-

handling the majority of the materials at the excavation sites. Materials that pose significant handling problems but which only constitute a small fraction of the total volume of material (e.g., intact drums) should be handled off-line at centralized facilities. These facilities would be located outside of the excavation sites so as to not inhibit excavation productivity. Another important objective of this approach would be to limit the generation of secondary wastes.

The system can be broken down into the following eight main subsystems and activities:

- Field screening and laboratory analyses
- Site containment
- Site excavation
- Oversized-object cutting
- Removal of pipelines under the water bodies
- On-site processing
- Waste packaging and transportation
- Site restoration

Definition of a field screening and laboratory analytical procedure is a critical linchpin to the approach.

Macroengineering places a strong emphasis on observational approach methodologies and real-time characterization as excavation proceeds, in contrast to extensive sampling and analysis prior to remediation activities. Consequently, a broad range of field detection capabilities should be developed. This includes real-time characterization for radiation levels, volatile organic compounds (VOCs), heavy metals, and anionic species.

To accomplish this objective, instrumentation packages mounted on telescoping booms should be utilized to characterize the working faces and bases of all excavations. Handheld and area-sampling equipment should also be provided for monitoring during excavation activities. Although these results would be useful in directing the work and alerting the operators to any imminent dangers, they would not be capable of providing absolute concentrations of most contaminants. As a result, samples would be collected periodically and analyzed in an on-site mobile laboratory. This would furnish the necessary cleanup confirmation data as well as provide a check of the field screening results. Approximately 10% of the mobile laboratory samples should be sent to a fixed laboratory that utilizes full quality-assurance (QA)/quality-control (QC) procedures as a further check on the mobile lab.

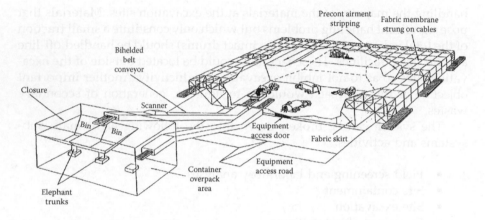

Figure 3.1 Mining-oriented macroengineering-containment structure.

- Containment structures such as a mobile truss system to prevent releases of contaminated dust to the environment
- Dust suppression measures such as water sprays, soil stabilizers, and vacuum hoods to control dust generation within the containment structures
- Ancillary support systems such as HEPA-filtered ventilation systems, fire suppression systems, primary and emergency power sources, and airlock entrances

Containment structures, such as that depicted in Figure 3.1, could be quite large and expensive to construct and operate. An attractive alternative at some sites may be the use of wind skirts to control fugitive dust emissions. In all cases, considerable engineering development will be required for the site containment systems prior to their implementation.

The mobile containment structure should have the following conceptual features:

- The capability for maintaining negative pressure inside the structure
- Be constructed of durable and reinforced structure material that can be decontaminated
- Be equipped with exhaust blowers, prefilters, and HEPA filters to remove contaminated particulates before discharging exhaust air to the environment
- Contain air codes to facilitate moving equipment and personnel

Chapter 3: Macroengineering technical approaches 27

Figure 3.2 Typical buckwheel excavations.

Large bulldozers and off-highway dump trucks could also be used during excavation and backfilling operations, such as that depicted in Figure 3.3. In areas where there is significant worker risk (such as radiation environments), control cabs on all equipment operating in contaminated areas should be fully enclosed, supplied with clean air, and shielded to meet As Low As Reasonably Achievable (ALARA) requirements. Excavated soil would then be transferred from the loader buckets to a conveyor system for transport out of the containment structure and into shipping containers. Large objects would be reduced in size prior to packaging and removal from the excavation site. Buried pipelines would be removed in a systematic fashion similar to that for soil.

If large objects such as steel and concrete structures and debris, timber cribs, and pipelines are encountered in the various waste sites, these items could require demolition and/or size reduction prior to packaging and removal from the excavation sites. Size reduction such as that shown in Figure 3.4 would only be performed to the extent necessary to facilitate handling and transportation. Hydraulically operated demolition and handling tools mounted on the booms of backhoes and excavators could be utilized for these purposes. The available tools (e.g., concrete pulverizer jaws, shear jaws, wood jaws, plate jaws, grapple jaws, and concrete cracking jaws) consist of a variety of powerful and interchangeable cutting, crushing, and grappling jaws; they can reduce almost any size and shape of steel or concrete to a manageable form.

Underwater pipelines present unique challenges and require special approaches when compared to land-based remediation activities. Due to the

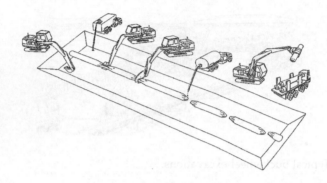

Figure 3.4 Size reduction — removal of buried pipeline on land.

potentially high costs of removal if these pipes and associated sediments are contaminated, precharacterization for contamination would be warranted. Remotely operated "moles" could initially be used to check for hot spots. All hot spots would then be sampled and analyzed for contaminant concentrations.

If the pipes and sediment contaminants are below cleanup standards, they could be excavated using barge-mounted cranes and clamshell excavators. If contaminant concentrations exceed clean-up standards, then sheet pile cofferdams could be constructed before excavation to contain contaminated sediments that would be stirred up during these operations. These sediments then would be placed into special shipping containers to allow dewatering prior to disposal. Water inside the cofferdams would then be analyzed and treated, if necessary, prior to removal of the sheet piles.

In keeping with the bulk-handling philosophy of the mining-oriented macroengineering technology approach, minimal waste processing should be conducted. However, some on-site processing operations would typically be required as outlined in the following.

Waste materials would be screened and segregated according to hazard levels for proper packaging and disposal. Size reduction would occur only to the extent required to facilitate handling, packaging, and transport. Waste volume reduction operations would be confined to field screening for contaminant levels during excavation; more complex activities, such as soil washing, would not be conducted.

Intact drums represent a special case. These should be set aside and analyzed off-line for VOCs. Drums that do not contain VOCs should be packaged for disposal. If VOCs are present, the drums should be sent to a centralized facility for thermal destruction prior to disposal of the residues.

Chapter 3: Macroengineering technical approaches 29

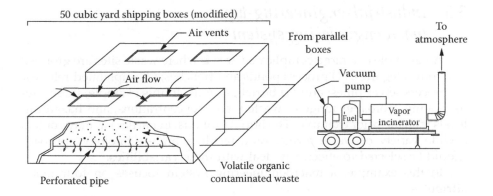

Figure 3.5 System for venting organics.

Relative to waste transportation, it is much more effective to utilize standard waste management containers because much of the cost in designing and constructing containers in the U.S. lies in the licensing aspect. However, if the site has special chemical or radioactive waste management issues and/ or has sufficient waste container volume requirements, specially designed configurations of steel containers could be utilized to transport waste forms. These configurations would be based on the waste particle size and any special hazardous characteristics. The containers should have special fittings for handling and securing during transport, depending on the type of transport.

Wastes could be transported to the disposal site on conventional bulkhead flatcars or trucks. Gantry cranes could be used to move the containers and pipes between the excavation sites and railheads or trucking points. In some cases, slurry pipeline could be utilized to efficiently move liquid and fine waste forms.

After a complete area has been remediated and certified clean, site restoration activities would consist of:

- Backfilling, compaction, and recontouring
- Topsoil spreading and preparation
- Reseeding and irrigation

In summary, the mining-engineering-oriented macroengineering system is based on an assumption that the site has a low complexity and sufficient size to warrant a bulk-handling approach. It stresses the fast-throughput approach, which utilizes minimal waste processing. The system should primarily use off-the-shelf equipment with some technology modifications and development

3.3 Industrial-engineering-type macroengineering system

An example of a more complicated area is where waste sites are grouped into four categories: (1) process ponds and trenches; (2) unplanned releases; (3) process sewer pipes; and (4) burial grounds. In this example, these sites have a much broader range of waste types, concentrations, and hazardous levels than the prior example. The burial grounds, in particular, may contain a wide variety of waste forms. As a result, a high-technology approach should be selected to effectively deal with these variabilities.

In this example, a macroengineering system focuses on four major attributes:

- Process flexibility to accommodate the wide variety of wastes that will be encountered
- Waste volume reduction to minimize the quantity of wastes shipped for disposal
- Equipment mobility to enhance its efficiency and reusability
- Proven technologies to accomplish the overall goals of the system

A goal of reducing the contaminated waste volume shipped will vary depending on the site, but a goal of reducing by 80% (as compared to the total volume of material excavated) is a reasonable goal to establish for a macroengineering cleanup. The first step in accomplishing this reduction is separation of the coarse and fine portions of the contaminated soils, thereby taking advantage of any known tendency of contaminants to concentrate in the fine particle fractions of site soils. Recycling and/or treatment of secondary wastes such as soil-washing fluids, decontamination waters, and organic contaminants could also be conducted to minimize the volume of waste materials sent to the disposal area. Additionally, some waste types could be shredded and supercompacted to reduce their volume prior to disposal.

Several differences exist between the general and industrial land-use options. For the industrial-use option, the lower cleanup standards would permit less soil excavation, reduced intensity of soil and debris washing, and reuse of some washed soils and debris as backfill in the waste-site excavations. The general-use scenario would require more soil removal than for industrial use; for example, more rigorous soil washing prior to use as backfill and disposal of all debris in the contained management area.

Similar to the mining-oriented system, the industrial-oriented macroengineering system should utilize field screening, mobile laboratory, and fixed lab analyses may also be needed to sort wastes for appropriate treatment, as well as to assess the effectiveness of the treatment processes.

Site containment systems would incorporate the same main features as those described in the earlier discussion on the mining-oriented approach. However, because of the large variations in waste-site sizes and shapes typical for industrial-engineering-oriented macroengineering sites, the more complex area may have to utilize a variety of flexible containment structure types, such as:

- Mobile bridge truss structures for large sites
- Frame-supported tents for small or long and narrow excavations

The tent structures could either be mounted on rails for mobility or will be disassembled and reassembled as appropriate. Wind skirts may also prove to be a viable alternative at some sites.

As for site excavation operations, such as the mining-oriented approach, the industrial-oriented approach should utilize large, conventional excavating equipment to remove soil and debris. Front-end loaders, backhoes, dozers, dump trucks, and the like would be used as appropriate for the given waste site. However, these machines would generally be smaller than their counterparts in the more mining-oriented system due to the higher variability in waste types at the complex industrial sites. Such variabilities, coupled with the need to minimize the volume of material that must be processed, would necessitate more precise excavation control, which is afforded by the smaller equipment. All equipment working in contaminated areas could be provided with supplied air and shielding for the operator cabs, if necessary.

Clean overburden soils should be removed to the extent practical before the site containment system is installed. For large waste sites, excavation of contaminated materials should be conducted in parallel panels (or cuts). This would be primarily to avoid the need for extremely large containment structures, because the rate of excavation will not justify such an enclosure over the entire site at any one time. Pipelines would be removed in a systematic fashion beginning at one end and continuing to the other.

If there is a tendency for hazardous constituents to adhere to finer fractions, then excavated soils should initially be wet-screened to separate and wash the material that is larger than 2 in. in size. This coarser soil fraction should be analyzed for hazardous constituents, and if it falls within clean closure standards, should be retained at the excavation site for future backfilling. The contaminated wash water and materials smaller than 2 in. could be transported via slurry pipeline to a soil-washing system.

For layers of soil with high levels of heavy-metal contamination

32   *Macroengineering: An environmental restoration management process*

using drum handling or grappling attachments on excavators or backhoes. Oversized objects should be reduced in size as needed to facilitate removal, packaging, and transport. Size reduction tools similar to those described earlier should be used for this purpose. The recovered objects should then be segregated at the excavation site according to activity level and material type (namely soils, drums, and other containers; metals, concrete, and other crushable debris; compressibles; and munitions). To simplify, further processing of soils should be wet-screened as described earlier, with the finer fraction pumped to the soil-washing system. The other materials should be placed into separate transfer containers for transport to the appropriate processing operation.

The industrial technology system may utilize a wide variety of processing and treatment methods to reach the goal of reducing the volume of waste to be disposed. This approach lends itself to both sequential and parallel multicomponent operation scenarios. The following is a brief discussion of examples of a multicomponent industrial system approach.

Intact drums and other containers should be processed to segregate the various material types contained in them. Bulging drums should first be vented in an explosion-proof atmosphere to remove their explosive potential. All containers could then be radiographed using real-time radiography (RTR). This would provide an initial screening of the containers' contents. If the containers are found not to contain explosives or other highly dangerous materials, they could then be transferred to a special glove box apparatus for further testing and waste segregation. The segregated wastes would then be treated in the appropriate systems described in the following.

Compressible and combustible wastes could be shredded in a rotary shear prior to further size reduction in a supercompactor. Shreddable metals could also be size-reduced in a series of rotary and rolling ring shredders. Shredded metals contaminated with organics could then be treated in the thermal desorption system. For the industrial-use option, the inorganic shredded metals could then be washed in the soil-washing system prior to use as backfill at the waste sites. For the general-use option, these metals could be supercompacted prior to disposal. Supercompactors can achieve volume reduction factors between two (2) and ten (10), depending on the material types.

Crushable solids (e.g., concrete, brick, clay pipe, and glass) could be processed through a series of jaw and rollercone crushers. Any steel, such as rebar, could be removed during this process and transferred to the shreddable metals system. The crushed materials would then be managed as described earlier for shredded metals. However, the supercompaction step would not be used.

Chapter 3: Macroengineering technical approaches   33

it is important to maintain the lowest potential for vaporizing metals and radionuclides. Both systems should be equipped with full emission controls.

In situ soil venting with vapor incineration could be used for all soils contaminated with VOCs.

Similar to the mining–engineering–oriented approach, after a complete area has been remediated and certified clean, site restoration activities should consist of backfilling, compaction, and recontouring; topsoil spreading and preparation; and reseeding and irrigation.

In circumstances in which munitions are encountered, the munitions should be placed into specially designed explosion-resistant containers and transferred to a U.S. Army demilitarization facility. If safe transport is not possible, military explosive experts should be called in to render the materials safe.

The slurried portion of contaminated soils from the excavations (smaller than 2 in.), as well as crushed and shredded debris under the industrial-use option, could be further treated by a soil-washing system. This system would further wash and separate the materials that are larger than 0.125 mm in size. Attrition scrubbing and acid washing would then remove the heavy metals and, if applicable, radionuclides, from this coarser soil fraction, and concentrate them in the finer fraction. The cleaned soils could then be returned to the excavations for use as backfill. The contaminated slurry would be dewatered prior to disposal. The resulting contaminated process waters would then be treated to remove the metals and, if necessary, radionuclides, and then recycled to the excavations and the soil-washing plant. The soil-washing process is a proven technology for remediating contaminated soils. Modular and transportable system components could be used to ensure versatility and efficiency.

In summary, the industrial approach meets the remediation needs when there is a higher concentration of wastes and greater complexity of waste mixtures. The proposed excavation and treatment system should use field-proven equipment from the mining industry and commercial-treatment industry; however, some engineering developments are still needed. Implementation of the proposed system includes the assumption of using waste volume reduction methods that yield a significant reduction in wastes requiring disposal.

Each of the technology orientations provides advantages and disadvantages that are to be evaluated when determining the ultimate engineering system to be implemented. On a strict process-cost-analysis basis, there is a considerable additional cost associated within an industrial approach vs. the mining approach on a cost-per-cubic-yard-processed basis.

However, other tradeoffs that must be considered are:

• The reduction in disposal costs associated with waste reduction

Taking into consideration the benefits of reducing disposal requirements, the industrial approach vs. mining approach technology tradeoff is reduced. Undoubtedly, final system designs could incorporate an efficient combination of both approaches if appropriate.

3.4   Contained management approach

A special element of the macroengineering application is where significant contamination from past practices precludes clean closure at a given area. In these situations, a contained management approach is more acceptable. The scope of the latter could typically include trenches, French drains, outfall structures, retention basins, sand filters, burial grounds, cribs, reverse wells, brine pits, ponds, ditches, diversion boxes, settling tanks, catch tanks, landfills, dumping areas, drain fields (selected), and unplanned releases. Other units that could be incorporated into this type of macroengineering restoration concept includes: storage tanks, pump stations, storage pads, ash pits, burning pits, control structures, neutralization tanks, evaporators, stacks, sewers, septic tanks, staging areas, or satellite areas.

Typically, for a contained management approach area, the area's mission has been so diverse that it is cost prohibitive to achieve clean closure without extraordinary technical developments and efforts. The mission of the engineering system in a large-scale restoration effort with this set of circumstances should be: (1) the development of proposed disposal sites for wastes recovered from other source areas (i.e., a hazardous waste management area), and (2) closing-in-place units that are not considered appropriate for clean closure.

However, it should be recognized that a contained management approach should be highly sensitive to community acceptance and surrounding community land-use goals. Without strong public outreach and acceptance, the contained management approach is always doomed to failure.

Closing-in-place can involve *in situ* stabilization of critical waste sites using technologies such as dynamic compaction, shown in Figure 3.6; and extracting any VOCs present using on-site soil venting methods depicted in Figure 3.7. New technology decisions for this set of circumstances should be developed with emphasis on both flexibility, given the variability of past practices at a given site, and implementability, given the long-term mission of the site.

The hazardous waste management area should be selected keeping in mind the following criteria to facilitate design and construction: It should (1) be sufficiently remote to surrounding community areas, (2) be easily controllable from an access standpoint, (3) have sufficient uncontaminated area to contain all of the retrieved wastes on the site, (4) have a relatively

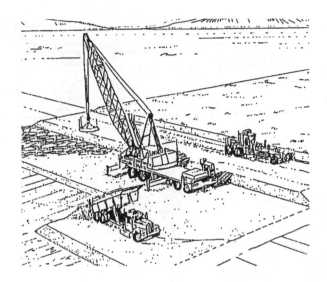

Figure 3.6 *In situ* stabilization and closure of a site.

processing, and disposal facility components. The receiving facility components must be designed to handle material shipped by either rail (both
hopper cars and flatbeds with containers), haul trucks, and, possibly, even
a slurry pipeline, as shown in Figure 3.8. Preferably, waste unloading activities should be conducted inside of enclosed buildings with airlocks, filtered
ventilation systems, and various decontamination systems. The processing
facility will typically provide for compaction or grouting of selected compressible material that pose potential subsidence problems and volume
reduction of soils. When special waste-handling issues occur (i.e., low-level,

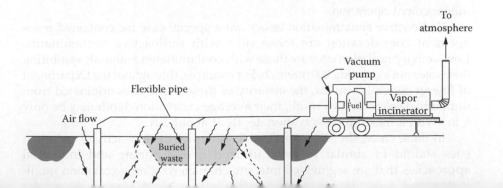

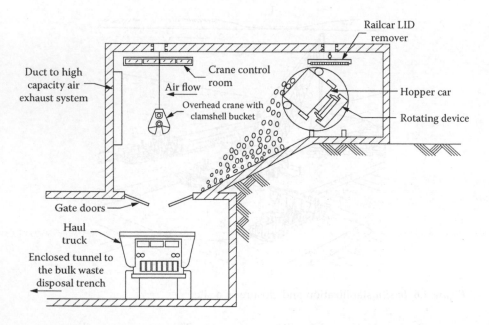

Figure 3.8 Railroad car unloading facility.

high-end radioactive wastes), separate unloading and disposal facilities should be provided and dedicated haul trucks should be available for transport of the received wastes. Containers, such as those containing high-activity radioactive wastes could then be lowered by crane directly from the unloading area into the disposal cells.

Trench systems are also a viable option for management of hazardous material. Figure 3.9 depicts a cross section of a trench hazardous waste management operation.

Radioactive contamination issues are a special case for contained management consideration are those sites with radioactive contaminants. Low-activity radiation sites are those with contaminated materials exhibiting dose rates not exceeding 200 mrem/h. For example, throughout the Department of Energy federal complex, the majority of these waste sites originated from unplanned releases. As a result, their average excavation depth may be only 6 in., with a maximum excavation depth of about 2 ft.

In these cases, the site investigation activities for low-activity radiation sites should be similar to those utilized in the mining and industrial approaches that are segmented into site characterization, excavation moni-

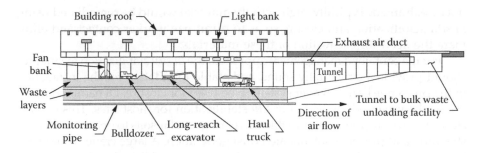

Figure 3.9 Trench hazardous waste management system.

as sagebrush, tumbleweeds, and other large plants could be removed using grappling equipment prior to soil excavation. These plants could then be baled and trucked to the disposal facility. For sites containing pipes, structures, and other large objects, backhoes and front-end loaders would be used to remove the obstructions. Size reduction and handling attachments could also be used to manage these objects. For sites in which special hazards occur (i.e., radiation, etc.), all equipment working in contaminated areas would be provided with supplied air and radiation shielding.

Whenever possible, a nonenclosed containment system is preferred for waste sites if it can be demonstrated that it provides adequate containment of contaminants. By doing so, the contractor:

1. Eliminates the need for decontamination of major structural facilities
2. Avoids the difficulty of moving an enclosing structure
3. Removes the problems associated with operating within an enclosed air space
4 Allows for faster, more cost-efficient remediation

An example of a nonenclosed containment system could include a system that combines spray-on dust suppressants, water sprays, wind breaks, and administrative controls to prevent contaminated dust migration. The system could provide an increased level of protection over ambient conditions to permit operations in both calm and moderate wind conditions. Operations would be shut down during excessive winds but would continue when meteorological conditions do not exceed the capabilities of the containment measures employed.

In special cases such as high-activity radiation contamination areas, the

38   Macroengineering: An environmental restoration management process

not be exhumed. Typically, high-activity material would be remediated using *in situ* stabilization and closure so as to minimize worker contact and eliminate the need for excavation of these materials.

For the contained management approach, the *in situ* stabilization techniques should provide a structurally sound and stable foundation for the installation of the protective barrier/cover system to be used for closure. Dynamic compaction is the preferred method of *in situ* stabilization for most sites. This process starts by covering the site with a thick layer of granular soils to provide radiation shielding and prevent contaminated dust generation. A large crane would then be used to repeatedly drop a heavy weight on a preestablished grid pattern. The force applied by the falling weight sends shock waves through the underlying materials that consolidate soil particles and debris, resulting in settlements and more densely packed materials. This is a fairly common and well-proven technology for construction and waste sites.

In some cases, pressure grouting might be required to fill voids that may collapse under the weight of the normal stabilization equipment, or cannot be crushed using dynamic compaction methods. A vibrating-beam technique can be used to inject the grout under pressure. This method is highly effective in producing a thoroughly grouted mass.

The barrier system to be utilized under a containment management approach should be constructed primarily of natural earthen materials to ensure its effectiveness over the long term. Figure 3.10 provides a schematic example of a barrier system. In this case, the barrier system consists of a base layer and side slope protection made of riprap. The latter acts as a physical barrier to large burrowing animals and discourages digging by intruding humans with hand tools. Layers of graded gravel and small pebbles is placed above the riprap and leveled. The pebble layer would then be covered with a geotextile (fabric) that inhibits small soil particles from filtering into the spaces between the pebbles. The final layer of soil should be a specific mixture of gravel, sand, and fine soil particles unique to the meteorological conditions of the site. The gravel prevents wind erosion. The sand and fine soil mixture is optimized to prevent excessive drainage or ponding of water. The latter mixture should be designed to encourage the growth of native plants. The action of plants extracting water from the soil, coupled with evaporation, should return essentially all of

Hanford protective barrier

the excess soil water back to the air. This would prevent infiltration of water into the underlying wastes, which could otherwise generate contaminated leachate and cause migration of contaminants into the surrounding soils and underlying groundwater.

3.5 Groundwater remediation approaches

The overall objectives that guide the design of the macroengineering-scale groundwater remediation systems are:

- Developing viable groundwater use and remediation options consistent with the future land-use alternatives and objectives
- Identifying existing groundwater remediation technologies that may be appropriate
- Developing sitewide engineering systems from the identified technologies that satisfy the restoration objectives for each scenario
- Identifying emerging technologies or research and development needs

Similarly, different options based on future groundwater/land use should be evaluated to demonstrate the effect of target objectives on the design and costs of the technical approach.

Engineering systems proposed for groundwater remediation are highly dependent on the ultimate time and cleanup objectives, just as land-use cleanup objectives impact cleanup for soil/source area removal operations. Engineering systems may involve direct cleanup via treatment systems and hydraulic control systems such as slurry wall construction (see Figure 3.11), use of hydrodynamic turbines, as well as aquifer injection and groundwater extraction systems using lixiviants to mobilize hazardous constituents and horizontal drilling methods.

When time is critical, more costly methods such as aquifer excavation with dredges may be used. Multiple scenarios should also be developed to evaluate future groundwater uses that require different degrees of technically sophisticated remediation. In this manner, both the site owner and the regulatory/public can effectively evaluate the cost/risk benefits relative to each scenario.

For illustrative purposes, Figure 3.12 presents an example of three sitewide remedial systems alternatives, one for each cleanup option.

As shown in Table 3.2, the groundwater scenarios can vary relative to the geographic point at which compliance to drinking water standards are met, as well as for the time frame during which the standards are to

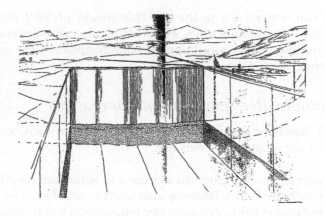

Figure 3.11 Slurry wall construction.

applicable in contained management restoration approaches, lixiviation enhances mobilization and extraction system effectiveness.

Fixation is a process that can improve groundwater quality by immobilizing the containment *in situ*. This process introduces a chemical compound that causes the contaminant to chemically bond and precipitate from the groundwater with the fixations. The fixation must be nontoxic and not degrade the aquifer, and its precipitated form with the contaminant should have low solubility and be geochemically stable.

Lixiviation, in contrast, enhances the dissolution of solid materials by introduction of a chemical agent. The objective in introducing lixiviating agents to the impacted aquifer is to enhance the rate of dissolution and, as such, decrease the amount of groundwater for treatment, which thereby decreases the time and ultimate cost of restoration.

Horizontal directional drilling methods have been used in the petroleum and utility industry for decades. From an environmental restoration

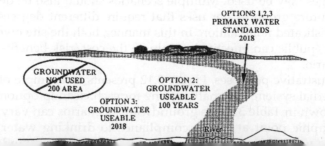

Chapter 3:   Macroengineering technical approaches   41

Table 3.2 Summary of Sitewide Groundwater Remediation Options

Study area	Technology orientation	Future-use objectives
Sitewide groundwater restoration approaches	Barrier technology used as nearest surface water source protection.	Option 1: Contaminant discharge to surface water is reduced. Sitewide institutional controls imposed and remain in place.
	Groundwater cleanup in long term; emphasis is on hydraulic control and current wastewater treatment technology.	Option 2: Contaminant discharge to surface water is reduced. Institutional controls and barriers are imposed to a limited area of groundwater. Remainder of site groundwater is restored to general use in unspecified long time frame.
	Groundwater cleanup within a definable term; emphasizes hydraulic control, aquifer mining, and state-of-the-art wastewater treatment with significant research and development required.	Option 3: Contaminant discharge to surface water is reduced. Institutional controls and barriers are imposed to a limited area groundwater. Remainder of site groundwater is restored.

perspective, these methods can be used in different ways. On the one level, horizontal drilling can be used to extract and intercept clean groundwater prior to movement through the affected area. Thus, the restoration program can isolate hydraulically the affected area and minimize downgradient impact.

On the other hand, the petroleum industry regularly uses horizontal wells for enhanced extraction from oil reservoirs and the same philosophy would hold true for certain (volatile-organics-dominated) environmental restriction efforts.

The options available in groundwater treatment systems include: ion

study efforts, but merely to provide a broad conceptual idea of the possible technical viability and the implications of a macroengineering approach to groundwater remediation. The engineered system for each scenario is intended to restore groundwater on a sitewide basis, rather than arbitrarily addressing groundwater restoration on an unrealistic individual-opera-ble-unit basis. Although aggregate areas may have distinct restoration systems, these systems should be designed to be compatible and integrated within each scenario for successful implementation and operation.

Bibliography

1. Barber-Greene, Co., *Telsmith Mineral Processing Handbook*, 7th ed., Telsmith Division, Milwaukee, WI, 1982.
2. Bauer, R.G., Containment Alternatives for Contaminated Soil Excavation, WHC-SD-EN-007, Rev. 0, Westinghouse Hanford Company, Richland, WA, 1991.
3. Bauer, R.G. and Fleming, D.M., Transportation Study for the Shipment of Contaminated Soils during Hanford Remediation, WHC-SD-EN-ES-009, Westinghouse Hanford Company, Richland, WA, 1991.
4. Caterpillar Inc., *Caterpillar Performance Handbook*, 19th ed., CAT Publication, Peoria, IL, 1988.
5. Chiarizia, R. and Horowitz, E.P., Study of uranium removal from groundwater by supported liquid membranes, *Solvent Extraction and Ion Exchange*, 8(1), 1990.
6. Conveyor Equipment Manufacturers Association, *Belt Conveyors for Bulk Materials*, 2nd ed., CBI Publishing Company, Boston, MA, 1979.
7. ENR, Slurry Wall Reaches New Depth, *Engineering News Report*, November 26, 1987.
8. Freeze, R.A. and Cherry, J.A., *Groundwater*, Prentice-Hall, Englewood Cliffs, NJ, 604 pp, 1979.
9. Gelhar, L.W., Mantoglou, A., Welty, C., and Reyfelt, K.R., A Review of Field Scale Physical Solute Transport Processes in Saturated and Unsaturated Porous Media, RP-2485-05, Electric Power Research Institute, Palo Alto, CA, 1985.
10. Golder Associates, *Groundwater Computer Package: User and Theory Manuals*, Golder Associates Inc., Redmond, WA, 1983.
11. Hay, W.W., *An Introduction to Transportation Engineering*, 2nd ed., John Wiley & Sons, New York, 1977.
12. Merritt, R.C., The Extractive Metallurgy of Uranium, Colorado School of Mines Research Institute, prepared under contract with the U.S. Atomic Energy Commission, Golden, CO, 1971.
13. Pfleider, E.P., *Surface Mining*, 1st ed., American Institute of Mining, Metallur-gical, and Petroleum Engineers, Inc., The Maple Press Company, York, PA, 1968.
14. Rexnord Inc., *Norberg Process Machinery Reference Manual*, 1st ed., Process Machinery Division, Milwaukee, WI, 1976.

17. United Nations, Physical Requirements of Transport Systems for Large Freight Containers, ST/ECA/170, United Nations Publication, Department of Economics and Social Affairs, New York, 1973.
18. WHC, Hanford Site Protective Barrier Development Program: Fiscal Year 1989 Highlights, Westinghouse Hanford Company, Richland, WA, 1990.
19. WHC, 100 Area Hanford Past Practice Site Cleanup and Restoration Conceptual Study, WHC-EP-0457, Draft, Westinghouse Hanford Company, Richland, WA, 1991.
20. WHC, 200 Area Hanford Past Practice Site Cleanup and Disposal Conceptual Study, WHC-EP-0454, Draft, Westinghouse Hanford Company, Richland, WA, 1991.
21. WHC, 300 Area Past Practice Site Cleanup and Restoration Conceptual Study, WHC-EP-0459, Draft, Westinghouse Hanford Company, Richland, WA, 1991.
22. WHC, Hanford Groundwater Waste Cleanup and Restoration Conceptual Study, WHC-EP-0458, Draft, Westinghouse Hanford Company, Richland, WA, 1991.
23. Chiarizia, R. and Horowitz, E.P., Study of Uranium Removal from Groundwater by Supported Liquid Membranes, *Solvent Extraction and Ion Exchange*, Vol. 8, No. 1, 1990.
24. Merritt, R.C., The Extractive Metallurgy of Uranium, Colorado School of Mines Research Institute, prepared under contract with the U.S. Atomic Energy Commission, Golden, CO, 576 pp, 1971.
25. Church, H.K., *Excavation Handbook*, McGraw-Hill, New York, pp. 13–151, 1981.
26. U.S. DOE, DOE Order 5480.3, Safety Requirements for the Packaging and Transportation of Hazardous Materials, Hazardous Substances, and Hazardous Wastes, Washington, D.C., 1985.
27. Buelt, J.L., Timmerman, C.L., Oma, K.H., and Fitzpatrick, V., *In Situ* Vitrification of Transuranic Waste: An Updated Systems Evaluation and Applications Assessment, PNL-4800 Suppl. 1, March 1987, Battelle Pacific Northwest Laboratory, Richland, WA, 1987.
28. Phillips, S.J. and Hinschberger, S.T., Concurrent *In Situ* Treatment and Disposal for Low-Level Radioactive Waste: Technology Development, Procedures and the International Radioactive Waste Management Conference, BNES, London 1989 2, 1989.
29. Gold, J.W., Compressed Gas Cylinder Disposal Alternatives, Superfund 1988 Proceedings of the 9th National Conference, Hazardous Materials Research Institute, Washington, D.C., pp. 183–187, 1988.
30. Taylor, D.W., *Fundamentals of Soil Mechanics*, John Wiley & Sons, New York, 1960.
31. U.S. DOE, Containment Alternatives for Contaminated Soil Excavation, CWHC-5D-EN-EV-007, Washington, D.C., 1991.
32. Gershey, E.L., Klein, R.C., Party, E., and Wilkerson, A., *Low-Level Radioactive Waste, from Cradle to Grave*, Van Nostrand Reinhold, New York, 1990.

chapter 4

Assessing innovative techniques and technologies

Because the macroengineering approach relies on innovation, it is recognized that successful implementation will require additional engineering developments and could benefit substantially from new technological breakthroughs. However, by viewing the environmental restoration mission in a "big-picture," "total system" manner, macroengineering provides a basis for identifying and prioritizing the technology development needs necessary to effectively complete a restoration mission. Therefore, it is critical to identify any potential gaps in technology and engineering developments and to determine if any of the gaps would impede the macroengineering approach.

A clear distinction is made between technology and engineering developments. A *technology development* requires putting significant research and development efforts into new and emerging applications that do not currently exist. It is estimated that several years would be necessary to bring these new developments to full-scale use. *Engineering developments* refer to adaptations or modifications of currently available equipment, systems, and technologies to meet unique program requirements.

Irrespective of whether one uses a mining, industrial, or contained management approach, every site cleanup or facility decontamination project should emphasize waste volume reduction and innovative waste handling technology options. These issues are overriding concerns in the selection of necessary cleanup technologies. Relative to environmental restoration operations, considerable success can be achieved through application of straightforward management techniques that emphasize control of secondary-waste generation. For example, an integrated process of collecting contaminated soil, sampling the soil, and verifying that the remaining soil is clean minimizes the volume of waste generated (by using a more "surgical" approach

special emphasis should be placed on avoiding the generation of any mixed wastes.

Relative to contained management approaches, technology development criteria may include defining the site stabilization parameters required to implement the program to an appropriate level for engineering specificity. For example, for dynamic compaction, the depths and levels of influence need to be determined for the different soil and waste conditions to be encountered. This involves establishing compaction weights, drop weights, and spacings.

The ability of grout to preclude subsidence relative to the concentrations of degradable wastes is another example of a technology needing field verification.

Likewise, the ability of *in situ* vitrification alternatives to stabilize degradable wastes needs to be evaluated via treatability testing relative to key operating parameters, such as melting rates, power demands, and off-gas generation rates and concentrations.

4.1   Independent technology survey

Many of the questions can be addressed not only with field testing, but also through an independent technology survey.

An independent technology survey involves surveying a wide range of vendor, supplier, and remedial action consultants to provide independent feedback on costs, schedules, and productivity issues that may impact the company's environmental restoration program. As part of the technology survey element, the researcher must garner independent data on technology performance relative to given technical, cost, schedule, and regulatory criterion, and just as importantly, identify "success criteria" as shown in Table 4.1, associated with identified and relevant environmental restoration technologies and approaches.

Other improvements that may not meet these criteria may still be deemed in the public interest and supported. Examples include actions that addressed public "perceived risk" attitudes and foster better land-use application above that deemed necessary for basic cleanup criteria.

Table 4.1 Environmental Restoration Management Audit

Technology success criteria

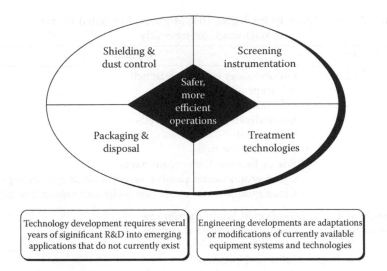

Figure 4.1 Macroengineering provides greater direction in identifying engineering and technology development opportunities.

As shown in Figure 4.1, much of the emphasis in engineering and development opportunities in macroengineering should be oriented towards issues that enhance safer and more efficient operations to the benefit of labor and worker safety.

An example of a straightforward operations process technology development investigation is a study to establish an acceptability types, frequency, and amount of dust suppressants and water spray to be used to minimize airborne containments during the excavation activities.

The technology survey element identifies engineering and technology development opportunities. Again, technology development opportunities being defined as requiring several years of significant research and development (R&D) into emerging applications that do not exist, whereas engineering developments are defined as adaptations or modifications of currently available equipment systems and technologies. Any successful environmental restoration program must be prepared to track and take advantage of both opportunities.

An example of the type of technologies that might be evaluated in a macroengineering study is listed in Table 4.2. Macroengineering projects require technologies and management techniques that focus on streamlining site characterization and remediation to allow for a safer and more expedi-

Table 4.2 Examples of Technologies That Might be Evaluated Under
a Macroengineering Study

Process	Option
Groundwater control	Groundwater extraction trench
	Intercept with reinjection
	Infiltration/recharge trench
	Groundwater extraction wells
	Slurry wall
	Damming the river
	Fox or lixiviate the contaminants
	Clean groundwater passive system buried gravity pipeline
	Clean groundwater extraction wells and subsurface pipeline
Groundwater treatment	Supported liquid membrane
	Granulated activated carbon
	Ion exchange
	Lime precipitation
	Reverse osmosis
	UV/peroxide, ozonation
	Air stripping
	Evaporation
	Microfiltration
	Biological denitrification
Rapid characterization	Cone penetrometer
	Photoionization detection of volatile organics
	Soil-gas analyses
	Metals analysis by X-ray fluorescence
	Geophysical techniques
	Directional (angle) drilling
	Mechanical moles
	Radioactivity survey
	Vertical drilling
Oversized object removal	Large pneumatic-tire mobile "picker" equipment
	Hydraulic excavation (Hydrex)
Oversight object treatment	Abrasive jet cutting
	Jaw crusher
	Rotary shear shredder
	Robotics
	Mobile shear
	Concrete pulverizer
Soil treatment	Vapor extraction
	Stabilization/solidification

Table 4.2 Examples of Technologies That Might be Evaluated Under
a Macroengineering Study (Continued)

Process	Option
Soil excavation	Continuous miners
	Cableway systems
	Dozers
	Draglines
	Hydraulic methods
	Hydraulic borehole mining
	Scrappers
	Shovels, trucks
	Rapid access tool (RAT)
Transportation	Conveyor
	Pipeline
	Rail
	Tram
	Truck
	Dust control
Dust control	Exterior containment
	Localized ventilation
	Moisturization water, foams, and fixatives
	Scrubbers — wet
	Electrostatic precipitation
Decontamination	High-pressure wash
Equipment shielding	Exposure/radionuclide screen
Personnel protection equipment	Respiratory air protection
Excavation and waste handling	Excavator
	Grappler
Radiation safety survey	Thin-window GM survey meter
	Alpha scintillation survey meter
	Fiddler probe
	Ionization chambers
	High-survey instrument
	Teletector
	Ambient radiation monitor
	Alpha air monitor

Table 4.3 Examples of Technology Development Opportunities That May be Applicable Under a Macroengineering Study

Recommended item	Recommended development or improvement	Long-term cost, schedule, or safety advantages
Real-time, analyte-specific quantification capability, e.g., concentrations of individual organic compounds and metals.	New analytical methods and/or detectors, both fixed and mobile.	Minimize excavation of soil that meets cleanup standards; no equipment standby time awaiting and analytical results from confirmatory sampling; lower cost analyses.
Field-screening instrumentation for radiation, chemical, physical, criticality detection.	Equipment made more mobile and less sensitive to adverse environmental conditions such as moisture, dust, vibration, and interferences.	Less equipment down-time because of lower maintenance/ replacement frequency; greater measurement accuracy and precision; increased safety assurance.
Chemical fixatives to control dust generation during excavation and material-handling operations.	Better penetrating, longer-lasting, more suppressive and nonhazardous chemicals.	Less need for expensive and restrictive containment structures; faster, more cost-effective excavations; safety level comparable to structures.
Lixiviants (and/or fixatives) for inorganic, metal, and radioactive containments in groundwater.	Chemicals that will readily release adsorbed (or permanently adsorb) contaminants during groundwater pump and treat options.	Reduced volume that is pumped and treated, resulting in lower costs, faster restoration of aquifer, lower residual aquifer concentrations for safer end use.

real-time screening, *in situ* monitoring instrumentation, and utilization of a state-of-the-art mobile laboratory to provide the majority of characterization data, and thus avoid unwarranted time delays. (The mobile lab data should

Chapter 4:   Assessing innovative techniques and technologies   *51*

Examples of Engineering Developments That Might be Applicable Under Macroengineering Study and Design and Operational Issues to be Addressed	Further design analysis	Modification of existing equipment or methods	Fabrication using existing materials	Concept performance testing	System optimization testing
...es		X			X
...eyors		X			X
...ation		X			X
...etectors	X				
...OC venting		X	X	X	X
...ure thermal				X	X
...ration	X	X		X	X
...ystem	X	X		X	X
...extraction systems			X		X
...and	X	X		X	X
...treatment			X		X
...l sheet pile					X
...nical	X				X
...monitoring at	X	X			X

52   *Macroengineering: An environmental restoration management process*

- Conducting a literature search to identify available technology methods and vendors
- Reviewing the documentation
- Verifying the laboratory and/or field performance data
- Identifying boundary conditions under which the technology is feasible from a cost, volume, and waste characteristic standpoint
- Defining the circumstances under which these methods might be applicable

A discussion of the technical approach and experience in reviewing incineration technologies is presented as an example of the information needed to evaluate the volume-reduction capabilities of a technology. The two key issues to be reviewed and controlled relative to waste volume reduction technologies, such as incineration, are: (1) waste acceptance criteria and (2) operation practices.

With respect to incineration, waste acceptance criteria are the key elements to establishing an effective quality control program to ensure that the operation is meeting emission limits and offsite exposures, as well as operating efficiency of the system. In the special case of radwaste, radiological characterization issues that need to be defined include:

- Acceptable and nonacceptable radionuclides
- Maximum allowed concentration for each acceptable radionuclide
- Maximum allowed quantity for each acceptable radionuclide
- Detailed characterization of the waste
- Nuclear criticality

A similar acceptance criteria must also be developed to address the hazardous waste characteristics (i.e., fluids, solvents, degreasers, lead, spent filters, and soil), both from a treatability and a corrosive property standpoint.

The experience of radioactive and mixed-waste incineration research, test, and evaluation is not as developed as it is for hazardous waste incineration. It has resulted in a lack of operational data to correlate incinerated waste characteristics and stack radionuclide emissions and to identify the suite of problems to consider for practical operation. However, typical problems to be expected and assessed are corrosion of components, plugging of heat exchangers, incomplete incineration, residual ash accumulation, off-gas system filter replacement, fires in the filter systems, HEPA filter clogging, humidity control, contamination control, generation of unacceptably high concentrations of radioactivity, radiation levels in ash (and accompanying

Laboratory testing:	Establish characterization protocols for the contaminated material and procedures for determining the applicability of various technologies to soil types.
Conceptual design:	Evaluate the field treatability systems that would include, for each system design, the equipment capability requirements, mass balance calculations from the contaminated soil spectra, the test parameters required, and the system design specification.
Bench scale testing:	Identify the physical techniques needed to identify particle behavior, develop test plans and procedures, and sampling plans and procedures. Test all pertinent operational equipment components, and develop process flow sheets that establish the production levels.

Figure 4.2 Evaluation of R&D volume reduction/waste minimization approaches.

sites of concern. Applicability of the technology to the full range of contaminant types and concentrations should also be evaluated. The need for using processing acids, such as surfactants, is another issue that should be investigated. Affiliated systems, such as slurry-characteristic requirements for the slurry pumping and piping system, should also be evaluated. Figure 4.2 identifies the specific elements of the assessment approach that the researcher may be called upon when evaluating soil-washing technologies. The focus of the first element, laboratory testing program support, would be to establish the envelope for which a technology is applicable based on contaminants and host materials. In the soil-washing example, soil characterization is the key parameter that controls the potential for removal of specific contaminants for both particle liberation volume reduction techniques and particle separation volume reduction techniques. The same can be said for other techniques such as chemical extraction, gravity separation, and magnetic separation. The focus of the second element, conceptual design review phase, is to establish whether the laboratory results generated for a given system design are implementable in the field. Issues such as preliminary system design specification system fabrication and assembly are reviewed for practicality and dependability. These issues are then given more stringent review during the bench-scale testing, which is designed to identify and evaluate problems prior to final specification of the field system.

Other examples of special plant-operation-testing requirements that may be associated with a macroengineering project include water treatment plant and grout plant development. Testing of water treatment technologies, dis-

For example, containers and lifting-frame designs should be tested and developed to increase the: (1) strength in design basis impact tests and (2) ease and security of grappling the container with a lifting frame. Transport vehicles enhancement could focus on seals and seal testing to verify leak tightness, ease of decontamination, and ease of maintenance. In addition, for vehicles in high-contamination zones, enhancements that minimize creation and dispersion of contaminated dust such as vehicle exhaust, engine intake, and radiator airflow improvements are attractive. Last, modeling the correlated field measurements to high precision, high-sensitivity laboratory chemical and radiological measurements would also add to worker safety, site operational efficiency, and control.

4.2   Technology risk assessment

A necessary element in assessing new technologies is the evaluation of its uncertainties and potential for failure. The question must also be asked, "How is the risk of failure in remediation to be measured?" Central to the measurement of the potential failure of a remediation program is the technology risk assessment (TRA). The latter can be defined in a systematic approach, identifying and evaluating risks associated with a remediation or a waste management technology. A TRA must consider risks associated with technology failures, indirect consequences, and primary and secondary risks of accidents and malfunctions. TRA does not duplicate or replace human health or ecological risk assessments but supplements it with the boundary condition of realistic technological expectations.

Each technology and treatment process evaluation is unique to the goals, materials, and technology options under consideration. The approach to evaluating the feasibility and cost of restoration options should consider the ability of each option to satisfy the demands of varying degrees of mitigation or cleanup criterion. Examples of environmental restoration control options could include the use of ion-exchange resins, soil washing, incineration, and simple disposal (recognizing EPA's preference for waste minimization and treatment over disposal). Evaluation criteria could include contaminant removal efficiency, capital cost, operating cost, and the potential for release of hazardous/radioactive constituents. The CERCLA remedy selection criteria constitute a useful basis for evaluating options. These criteria include:

- Overall protection of human health and the environment
- Compliance with ARARS
- Long-term effectiveness and permanence

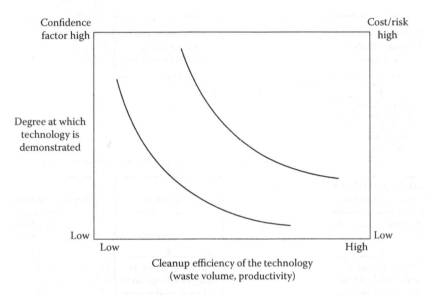

Figure 4.3 Technology risk assessment.

The reviewer should consider these and other criteria identified as appropriate in its evaluations of control options and strategies.

Figure 4.3 presents a schematic defining the universe of technology risks — primarily centered on a given technology's design and functionality characteristics. All of these potential risks should be considered to determine the extent to which technology risk impacts the environment and drives costs.

As depicted in Figure 4.4, identification of environmental restoration control options should be followed with a screening of available option steps to identify those technology options that are likely to justify in-depth evaluations and investigations. The screening activity will consider the performance of each potential option relative only to the criteria judged to be most significant. In other words, the screening process may consider the performance of options relative to only a few of the nine criteria listed earlier. The significant criteria include those involving public health and the environment and costs.

Once the suite of probable options (including combinations of options) has been identified, the reviewer should undertake a parallel series of activities that seek to characterize performance of each option in terms of all evaluation criteria, as shown in Figure 4.4. Implicit in the exhibit is the assumption that performance, in terms of limiting effects on public health

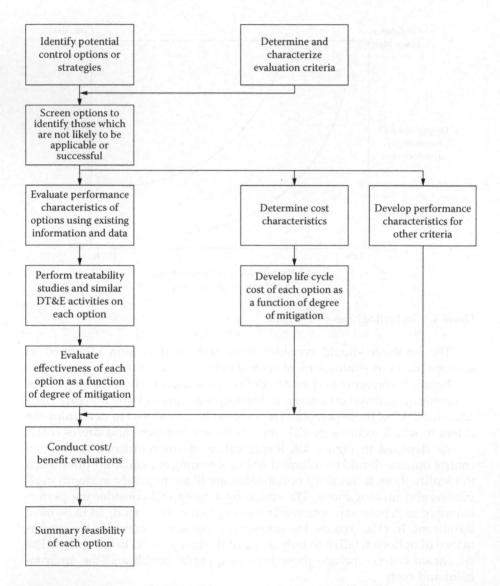

Figure 4.4 Sequence of activities to evaluate control options and strategies for varying degrees of mitigation or cleanup.

or desirable, the reviewer should conduct laboratory-, bench-, or pilot-scale treatability and other developmental test and evaluation (DT&E) investigations to develop an adequate database of performance information

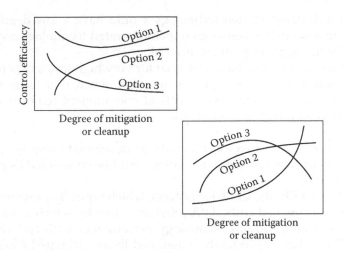

Figure 4.5 Example of how performance might be portrayed as functions of the degree of mitigation or cleanup.

reviewer must determine the efficiency with which contaminant-release potential and risk to public health and the environment are controlled, as functions of the extent to which contaminant must be removed, stabilized, or isolated.

The reviewer may summarize the results of these evaluations in a form similar to that portrayed in Figure 4.5. As shown earlier in Figure 4.4, the reviewer should pursue other evaluations parallel with those described earlier. These evaluations will address performance of the control options in terms of the other remedy-evaluation criteria. As noted earlier, cost factors will be important considerations, and should always be evaluated in conjunction with the efficiency and public risk tradeoffs described earlier. Cost data will be taken from information provided by vendors, cost guide data (e.g., the environmental cost handling options and solutions (ECHOS) environmental restoration unit costs), and historical data.

The reviewer should also add TRA to the CERCLA remedy selection criteria. TRA provides a mechanism for considering site- and implementing-party-specific considerations into the selection process. The TRA criterion basically encompasses an analysis of cost and technology performance risks that are not considered in the CERCLA remedy selection criteria. The following are discussions of these TRA considerations:

- A well-demonstrated technology would have a low degree of performance risk, whereas an undemonstrated technology would have a high degree of performance risk.
- A technology that has a potential for very high levels of contaminant removal may be assigned a low risk, whereas a technology that achieves relatively lower levels of contaminant removal would be assigned a higher level of risk.

A composite performance risk could be developed using the preceding information, or each risk characterization could be considered as part of the decision process.

Cost risk: Unlike the CERCLA process, which typically provides a simple statement of estimated costs, TRA provides a tool by which a party implementing a cleanup can link technology performance with cost risk. In the case of TRA, a level of certainty is assigned to the estimated costs for each control option under consideration. Levels of certainty are also assigned to the levels of performance that could be achieved by each control technology. For example, a technology that could achieve 50% contaminant removal under virtually all conditions would have a relatively assured total (initial and operating) cost. This cost could be assigned a lower level of risk. Continuing this example, if some data show that the same technology could achieve 75% contamination removal, albeit on a less assured/consistent basis, total costs could be reduced, but a high level of risk would be assigned.

In combining these risk considerations, the cost of risk is based on an organization's attitude towards risk. For example, a risk-averse entity may choose a control technology that is well demonstrated, but which may be more costly (based on initial costs and/or long-term costs). In another case, an organization that may be willing to assume a greater level of risk may select a technology that has less or no history based on an expectation of lower costs and more rapid cleanup.

Implementing TRA would not diminish the role of regulators or of the community. Instead, TRA provides additional data to explain in greater detail the options available to the implementing entity, the regulators, and the community, and to refine the decision-making process.

Bibliography

1. Baker, E.G., Pohl, P.T., Gerber, M.A., and Riomath, W.F., Soil Washing: A Preliminary Assessment of its Applicability to Hanford, WA, U.S. Department of Energy Contract DE-AC06-76RLO 1830, 1990.

Chapter 4:   Assessing innovative techniques and technologies   59

4. Louisiana Tech. Engineer, Horizontal Directional Drilling/Guided Boring Methods, Louisiana Tech University, Vol. 45, No. 1, 3–11, Fall 1990.
5. Ricketon, W.A. and Boyle, R.J., Solvent Extraction with Organophosphenes Commercial and Potential Applications, *Separation Science and Technology*, Vol. 23, No. 12,13, 1988.
6. U.S. EPA, The Superfund Innovation Technology Evaluation Program: Technology Profiles, EPA/540/5-90/006, Washington, D.C., 1990.
7. U.S. EPA, The Superfund Innovation Technology Evaluation Program: Terra Vac *In Situ* Vacuum Extraction System, EPA/540/A5-99/003, Washington, D.C., 1989.
8. U.S. EPA, Assessment of International Technologies for Superfund Applications, EPA/540/2-88/003, Washington, D.C., 1988.
9. U.S. EPA, Superfund Treatment Technologies: A Vendor Inventory, EPA/540/2-86/004(f), Office of Solid Waste and Emergency Response, Washington, D.C., 1986.
10. Esposito, P., Hessling, J., Locke, B.B., Taylor, M., Szabo, M., Thurnau, T., Rogers, C., Traver, R., and Barth, E., Results of treatment evaluations of contaminated synthetic soil, Reprinted from *JAPCA Journal*, Vol 39, No. 3. 1989.
11. HMCRI, Proceedings, National Research and Development Conference on the Control of Hazardous Materials, Sampling and Monitoring, New Orleans, LA, 1991.
12. *Journal of air and Pollution Control Association* (now Called *Journal of air and Waste Management Association*) (JAPCA), Results of Treatment Evaluations of a Contaminated Synthetic Soil, Vol. 39, No. 3. 1991.
13. Richardson, W.S., Hudson, T.B., Wood, J.G., and Phillips, C.R., 1989. Characterization and Washing Studies on Radionuclide Contaminated Soils, Proceedings of Superfund 1989, 10th National Conference, Washington, DC, pp. 198–201.
14. U.S. EPA, U.S. Air Force, Remediation Technologies Screening Matrix and Reference Guide, Emergency Response (OS-110W), Washington, D.C., July 1993.
15. U.S. DOE, Office of Environmental Management, Office of Technology Development. VOCs in Non-Arid Soils Integrated Demonstration, Technology Summary, Washington, D.C., February 1994.
16. U.S. DOE, Office of Environmental Management, Office of Technology Development, VOCs in Arid Soil, Technology Summary, Washington, D.C., February 1994.
17. U.S. DOE, Office of Environmental Management, Office of Technology Development, Robotics Technology Development Program, Technology Summary, Washington, D.C., February 1994.
18. U.S. DOE, Office of Environmental Management, Office of Technology Development, Mixed Waste Landfill Integrated Demonstration, Technology Summary, Washington, D.C., February 1994.
19. U.S. DOE, Office of Environmental Management, Office of Technology Development, *In Situ* Remediation Integrated Program, Technology Summary, Washington, D.C., February 1994.
20. U.S. DOE, Office of Environmental Management, Office of Technology Development, Supercritical Water Oxidation Program (SCWOP), Technology Summary, Washington, D.C., February 1994.
21. U.S. DOE, Office of Environmental Management, Office of Technology

chapter 5

Site characterization

Site characterization can be broken down into two parts:

1. Preremediation characterization activities that encompass both defining the unique natural conditions of the site (groundwater, surface water, geology, soil, climate, etc.) and the site environmental insult/waste management history (type and location of waste management units and spills, waste types handled, etc.).
2. The characterization activities that occur as part of the environmental remediation process. The latter emphasizes waste characterization and confirmation of original assumptions in regard to the type of waste, extent, etc.

Under macroengineering, site and waste characterization becomes the means of defining the general nature and extent of contamination. The overall objective of site and waste characterization activities is to obtain only such information as is necessary to allow comparison of remedial action alternatives for the selection of remedy during the feasibility study. Special considerations are extended in macroengineering to include the following additional factors: protect workers during remediation, minimize environmental impacts, avoid unnecessary waste volume expansion through unnecessary excavation outside the boundaries of contaminated areas of concern, provide design input parameters for assessing treatment and handling options, as well as monitor remediation effectiveness. A heavy emphasis will be placed on nonintrusive site characterization techniques and the use of mobile screening laboratory facilities to facilitate remediation characterization control. As shown in Figure 5.1, site characterization entails identifying and characterizing all contaminant migration pathways from solid-waste management units (SWMUs). The objectives of site characterization programs will vary depending on the technical options being considered by the environmental restoration program. Figure 5.2 provides an example of the

62   *Macroengineering: An environmental restoration management process*

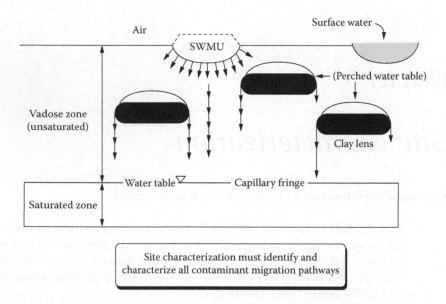

Figure 5.1 Site characterization goal — identify all contaminant migration pathways.

As shown in Table 5.1., there is a wide array of alternatives for site investigations, ranging from intrusive (geotechnical-oriented) methods to nonintrusion (geophysical- and geoprobe-oriented) methods. Likewise, there is a wide array of options relative to the associated issues such as radiological, unexploded ordnance (UXO), and chemical techniques. Macroengineering is biased towards utilizing the nonintrusive approaches as much as possible, and that is the emphasis of the site characterization methods discussed in this chapter. The more conventional geotechnical methods are well established within existing ASTM standards and should be referenced by investigators as necessary.

5.1   Hydrogeologic investigations

In the macroengineering example, we anticipated the need for observation well samples (taken before closeout as the groundwater level stabilizes). However, the latter is time consuming and expensive, and should be minimized to the number of wells needed to adequately determine and track site water quality. For one thing, drilling spoils must be disposed of as hazardous or radioactive waste. Secondly, drilling methodologies are inherently disruptive, and test wells often serve as a pathway for groundwater contamination to occur when slug-tested and to a greater degree if pumped in drawdown

	Drains & trenches	Capping	Slurry wall	Runoff/ run-on control	Gas ventings	Solidifi-cation	*In situ* soil flushing	Biore-media-tion	Vacuum extraction	Pump & treat
er table				'						
/groundwater		'		'	'	'		'	'	
flow rates &		'	'	'	'	'		'	'	
nges in elevation			'	'						
nductivity/				'		'				'
ipitation			'		'		'			
characteristics				'						
concentration		'		'						
tamination				'						
nesses, and extent & unsaturated aterials		'		'						
ristics		'				'				
ntent	'	'	'	'			'			'
permeable zone	'	'	'	'	'	'	'	'	'	'
itard	'			'	'	'	'	'	'	'

characterization data needs for various technology options.

...nary of Site Investigation Methods

	Hazardous waste areas	High-activity radiation areas	UXO areas
...ploration	Unmanned survey vehicles Computer-aided design	Unmanned survey vehicles Computer-aided design	Unmanned survey vehicles Computer-aided design
	Ground-penetrating radar EM induction Driven electrode conductance Passive metal detector Cone penetrometer	Ground-penetrating radar EM induction Driven electrode conductance Passive metal detector Cone penetrometer Seismic refraction	Ground-penetrating radar EM induction Driven electrode conductance
Initial:	Shovel/trowel Grain sampling tube Sampling tube Electro roto hammer Hand auger/post hole digger Hammer-driven ram pipe/split spoon	Not appropriate due to high level of radioactivity	Not appropriate due to potential UXO activity
Advanced:	Power auger Truck-mounted hydraulic ram pipe Backhoe Truck/trailer-mounted auger drill	Advanced: Power auger Truck-mounted hydraulic ram pipe Backhoe Cable drill	
...alysis			

Chapter 5:   Site characterization   65

sis

Wet chemistry
Ion selective electrode
Soil conductivity
Monoclonal antibody test
Drager tubes x-ray fluorescence
Field portable gas chromatograph

Panel or oversized van
Commercial truck/trailer
Gas chromatograph coupled with:
Flame ionization detector
Electron capture detector
Flame photometric detector
Nitrogen-phosphorous detector
Electrolytic conductivity detector
Helium plasma elemental detector
Mass spectrometer
Flame atomic adsorption
Graphite furnace atomic absorbance
Cold vapor atomic absorbance
x-ray fluorescence

Conventual laboratory capable of
handling hazardous waste samples

Radiation badges
Gamma meters
Beta/gamma meters
Alpha/beta/gamma meters

Conventional radiological
laboratory located in area

Not appropriate due to the high level
of radioactivity associated with
collected samples

Not appropriate due to high level
of radioactivity associated with
collected samples

Conventional laboratory capable of
handling radioactive samples

EOD personnel on-site

Wet chemistry
Ion selective electrode
Soil conductivity
Monoclonal antibody test
Drager tubes x-ray fluorescence
Field portable gas chromatograph

Panel or oversized van
Commercial truck/trailer
Gas chromatograph coupled with:
Flame ionization detector
Electron capture detector
Flame photometric detector
Nitrogen-phosphorous detector
Electrolytic conductivity detector
Helium plasma elemental detector
Mass spectrometer
Flame atomic adsorption
Graphite furnace atomic absorbance
Cold vapor atomic absorbance
x-ray fluorescence

Conventional laboratory capable
of handling radioactive samples

66   *Macroengineering: An environmental restoration management process*

air and hydraulic-driven systems provide advantages in speed, they also provide the most potential for intrusive disruption and associated release. Although cable tool methods are less destructive and offer more control, they are also slower and potentially tortuous in some soil/rock environments.

Thus, a network of wells must be developed that are sufficient to define the currently existing plume of contamination. The data must be sufficient to define critical aspects of the contaminant migration, including vertical and horizontal extent of the contaminant plume, temporal trends in the behavior of the plume, presence of preferential pathways, and the effect that local surface water bodies may have on migration rates and directions.

In short, the two basic objectives of the groundwater monitoring program are to:

1. Provide sufficient information from which both upgradient and downgradient concentrations in groundwater can be established, as well as estimate the overall mass loading rates to surrounding surface water bodies.
2. Assess the degree to which contaminants are confined to the site area and detect any statistically significant releases prior to contaminants migrating off-site. There is also a need to identify the plume centerline, as well as the transient behavior of the plume. Clearly, data needs must be balanced against the overall objectives and the practicality of performing long-term monitoring of numerous wells.

Geoprobe sampling can be an alternative to numerous monitoring wells. Geoprobes can be used to both define the plume and precisely locate monitoring wells, thereby reducing the number of wells. Recognize that data from exploratory drilling will provide only a snapshot of the distribution of contaminants. The latter will probably change over an annual cycle and with year-to-year variations in recharge precipitation. As such, there are a number of advantages to collecting hydropunch samples to optimize the location of future monitoring wells. These advantages include the following:

*   Concentrations can be far more representative of actual pore-water values, as they are not integrated over the entire screened interval
*   Hydropunch samples provide a much better three-dimensional view of the plume morphology
*   Hydropunch samples facilitate the identification pathways

Note that although the concentrations will change with time, it is likely a numerical modeling study can be performed to optimize the placement of the monitoring wells. Factors that need to be considered in the modeling include plume dispersion, contaminant velocity, nature of the source release (e.g., pulse or continuous), preferential pathways, and sampling frequency. A number of computer programs are designed specifically for this purpose. Also, a monitoring plan should be developed so that statistical comparisons can be made to determine whether there are any significant differences in chemical and/or radionuclide concentration between downgradient and upgradient monitoring wells.

The goal of the program is to have sufficient data to develop a numerical model of the groundwater flow system that can be used to obtain reliable estimates of dilution, groundwater flow rates, and groundwater flow directions. The contaminant transport model depends on two fundamental sets of data — hydrogeology and water-aquifer chemical reaction (chemistry of attenuation). The hydrogeologic and transport model must also be geared towards addressing a variety of uncertainties — source area loading; surface water infiltration; groundwater–surface water interaction; seasonal variations; sensitivity analyses for dispersivity; flow rate; hydraulic conductivity; homogeneous vs. heterogeneous nature of the aquifer (i.e., channeling?); hydrologic interaction; contaminant attenuation characteristics; groundwater usage impact on-site hydrology/ contaminant plume; and impact of catastrophic events (i.e., floods).

5.2 Soil and sediment investigations

Surface soil and near-surface soil are collected before, during, and after the remedial activities for several different purposes. Sampling plans must be developed that specify the numbers, locations, purpose, and rationale for collecting each type of soil sample.

During the actual remedial action program, soil samples will be collected and analyzed for four different purposes.

The first set of material samples are collected during the remedial activities and are referred to as *material characterization samples*. This group includes samples that are collected and analyzed during the course of the soil excavation activities. These samples will be collected from the excavation floor foundation material and from elsewhere on the site (e.g., loadout area) where surveys and preremedial action characterizations show levels to be at the threshold values of concern. The collection of these samples and rapid analyses in the on-site laboratory provide a means to quickly assess whether or not soil materials should be sent off-site for disposal.

For example, if soils are encountered that appear to contain organic

collected from the pile and analyzed in the on-site lab. This information will be used to fill out the shipping papers.

The third type of soil samples are those used to characterize the organic contaminants for the disposal/reprocessing facility's permits. Split samples are typically collected from the composite samples collected to characterize the concentration for shipping and/or submitted to the off-site lab for VOCs, SVOCs, metals, pesticides, and PCBs.

The fourth set of soil samples to be collected during the remedial project is termed the *final status survey samples*. These samples are collected from excavation pit floors and side slopes after all contaminated soils have been removed and a walkover survey of each excavation area has been performed. They will also be collected from other survey units around the site. The purpose of these samples is to confirm that all contaminated soils exceeding the clean up criteria have been removed. These samples are typically sent to the off-site laboratory for analyses. Details of the statistical sampling design, sampling procedures, numbers of samples, and analytical protocols should be described in a final status survey plan.

Sediments that are the result of surface run-off or decontamination procedures should also be sampled during the course of the project. These potentially contaminated samples may involve sediments that accumulate in sumps, water retention structures, natural drawings areas, and the silt fences. Each of these sediment locations should be sampled and analyzed based on sediment build-up in the field laboratory. When necessary, the sediment that has accumulated at these locations should be removed. The laboratory analyses will dictate whether the sediments need to be placed in the clean soil storage pile, or whether it should be placed in the pile destined for off-site disposal/recycle.

All soil and sediment samples collected during the project should be discrete samples, except for the composite soil samples that will be collected from the material that is destined for off-site disposal.

5.3   Data quality objectives

The intent of soil/sediments/ground-density-sampling efforts is to characterize material sufficiently to ensure a within-95% confidence level (±2 standard deviations of laboratory measurements) that the results are within the selected criteria for the proper disposition of the material. Samples are also typically collected to monitor site conditions in accordance with federal and state regulations. The data results will be used to determine the disposal path of the

excavated). *In situ* sampling is performed for material profiling or for material acceptance criteria prior to excavation. *In situ* sampling may also be performed to verify the presence of chemical contamination, as indicated by screening results obtained in the field.

The data quality objective (DQO) statements should be designed to address the data requirements. These statements should include the following elements:

- Intended uses of the data:
  Does the data satisfy the project objectives?
- Data need requirements:
  Data user perspective (i.e., risk, compliance, remedy, or responsibility) satisfied
  Contaminant or characteristic of interest identified
  Media of interest identified
  Required sampling areas or locations and depths identified
  Number of samples required (fixed number or dynamic estimate; probabilistic or nonprobabilistic basis)
  Reference concentration of interest or other performance criteria (e.g., action level, compliance standard, decision level, and design tolerance) identified
- Appropriate sampling and analysis methods

  Sampling method (e.g., discrete or composite sample; sampling equipment and technique; quality assurance/quality control samples) identified
  Analytical method (e.g., sample preparation, laboratory analysis, method detection limit and quantification limit; and laboratory quality assurance/quality control) identified

The development of the DQO statements should involve the entire project planning team. This ensures that all concerns from all of the project disciplines are addressed. To ensure that the data results meet the project objectives, the sampling program and the analytical methods chosen must meet PARCC requirements (PARCC stands for precision, accuracy, representativeness, comparability, and completeness). The following discusses how each requirement will be achieved for sampling and analytical methods.

5.3.1 Precision

70   *Macroengineering: An environmental restoration management process*

and processing procedures and by analyzing matrix duplicate and matrix spike duplicates.

Precision is calculated by relative percent difference (RPD):

$$RPD = \frac{\text{Difference between two measured values}}{\text{Average of the two measured values}} \times 100$$

If sufficient duplicate samples are collected (minimum of eight), then precision can be calculated by relative standard deviation (RSD) or coefficient of variation (CV). This value looks at the precision of the sample population from within the matrix. The formula to calculate RSD is as follows:

$$RSD = CV = (\sigma\sqrt{x}) \times 100\%$$

$$x = \left\{ \frac{\sum (x_1 - \bar{X})^2}{(n-1)} \right\}^{\frac{1}{2}}$$

where σ = population standard deviation.

5.3.2   Accuracy

Accuracy is the measurement of the difference between actual measurement and the true or expected value.

Accuracy from an analytical perspective is achieved by preparing and analyzing laboratory control samples and method blank samples. These will be analyzed on a frequency of one sample per matrix batch.

Accuracy is assessed as a percent recovery (R) or as a percent bias (R – 100). Percent recovery is calculated by the following formula:

$$\frac{(x_s - x_u)}{K} \times 100\%$$

where x_s = measured value of the spiked sample, x_u = measured value of the spiked sample, and K = known amount of the spike in the sample.

established standard operating procedures (SOPs) and monitoring sampling activities. Proper sampling techniques and proper homogenization techniques achieve this. Analytically, the aliquot obtained from each sample must be representative of the original sample, thereby ensuring quality results.

5.3.4   Comparability

Comparability is achieved in the field and in the laboratory by the use of established SOPs for sample collection, analysis, calculations, and reporting.

5.3.5   Completeness

Completeness is a percentage of usable results vs. the total number of results obtained. At a maximum, the goal for completeness is 100%.

Completeness may be calculated for the project as follows:

$$\%C = \frac{V}{N} \times 100\%$$

where V = number of measurements judged valid and N = number of valid measurements needed to achieve a specified statistical level of confidence.

5.4   Geophysical investigations

Subsurface detection, location, and characterization of UXO, hazardous waste, radwaste and Brownfield sites is a complex undertaking that requires the application of extensive experience in multiple disciplines. Subsurface mapping also requires knowledge of site-specific conditions that effect data acquisition and interpretation. The wide range of possible subsurface targets and structures existing at these sites requires a comprehensive approach to the problem. For example, subsurface UXO targets can be made of ferrous or nonferrous metals and the efficient characterization of these subsurface items requires the use of different sensor technologies for their accurate, rapid, and cost-effective mapping. Additionally, the unique history of UXO, hazardous waste, radwaste, and Brownfield sites underscores the importance of incorporating historical, planimetric, and archival information into

72   *Macroengineering: An environmental restoration management process*

Ground penetrating radar (GPR)
Responds to changes in electrical properties which are
a function of soil and rock material and moisture content.

Electromagnetics (EM)
Measures bulk electrical conductivity which is a function
of the soil and rock matrix, percentage of saturation, and
type of poor fluids.

Magnetometry (MAG)
• Detects ferrous metal (iron or steel) only.
• Response is a function of the object's depth and mass.

Portable organic vapor analyzers (OVA/GC)
• Measures organic vapors to levels of 0.5 ppm.
• Response is a function of organic's volatility and quantity.

Figure 5.3 Contemporary site investigation techniques, part 1.

(1) geophysical sensors, (2) global positioning system (GPS) navigation,
(3) geographical information systems (GIS), and (4) geophysical data analysis.

There is a wide array of alternatives for site investigations ranging from
intrusion to nonintrusion methods. Likewise, there is wide array of options
relative to associated issues such as radiological, UXO, and chemical analytic
techniques and statistical sampling methods. Nonintrusion survey methods
offer less risk while still providing a great deal of waste unit, waste type,
and contaminant boundary information. However, nonintrusive survey
methods still require a level of ground truthing with hard intrusion data.
Several geophysical sensors that are frequently used for subsurface characterization are listed in Figure 5.3 and discussed further in the following text.
Figure 5.4 presents an evaluation of a contemporary suite of site investigation
techniques.

Surface geophysical surveys can significantly reduce intrusive testing
and costly analytical work. Besides having a cost savings impact, when
properly planned, executed, and interpreted, surface geophysical surveys
can expedite site characterization and remediation activities. Surface geophysical surveys can be used to target areas of concern and be the basis for

P— *Indicates the primary choice under most field conditions*
S— *Indicates the secondary choice under most field conditions*
L— *Indicates limited field application under most field conditions*
A— *Not Applicable*

echniques

Geophysical

	GPR	EM
AG		
ocation of Buried Wastes and Delineation Trench Boundaries:		
- Bulk Waste Trenches	P	P NA
- Bulk Waste Trenches	P	P P
- Depth of Trenches and Landfills	S	L NA
- Detection of 55-Gallon Steel Drums	S	S P
- Estimates of Depth and Quantity of 55-Gallon Steel Drums	S	L P

.4 Contemporary site investigation techniques, part 2.

74   *Macroengineering: An environmental restoration management process*

given site situation. Background values can influence data quality due to a number of natural causes, such as variation in soil types, depths of overburden, as well as elevation differences. The recognition of background trends, coupled with the selection of optimum display formats, can significantly increase the investigator's ability to differentiate between background and anomalous features due to buried waste, for example.

Surface geophysical surveys have been used to detect buried ferromagnetic debris, drums, USTs, pipes, landfills, sludge lagoons, uncontrolled waste pits, trenches, extent of contaminated groundwater plumes, leachate plumes, buried (abandoned) utilities, voids, and old underground mine workings.

Ground-penetrating radar (GPR) responds to changes in electrical properties that are a function of soil and rock material and moisture content. GPR provides a continuous picture-like display that can be effectively used to locate metal objects, such as waste containers or burial trenches in the subsurface, or for profiling subsurface soils. GPR provides very high resolution, but its depth of penetration is site specific. Penetration can be as great as 100 ft but is commonly less than 30 ft. Potential problems can occur if large underground metal structures are nearby. Without some type of filtering, the latter may interfere with depth performance and mask anomalies. Current systems are designed to be towed by hand or on a sled vehicle. Traverse speeds may vary from 0.5 to 5 mi/h.

When targets are made of nonferromagnetic materials (copper or aluminum, for example), magnetometers and/or gradiometers are inadequate for detection. In these cases, electromagnetic (EM) sensors that exploit pulsed induction phenomena should be used. EM induction can be used to locate areas of differing conductivity below surface. EM measures bulk electrical conductivity, which is a function of the soil and rock matrix, percentage of saturation, and type of pore fluid. Drive electrode conductance is determined by driving electrodes into the soil, applying a voltage and measuring current. The system would measure the conductive difference of soils. Depth of insertion and spacing of the electrodes would determine the depth of the survey. EM sensors can be hand carried or vehicle mounted (preferably nonmetallic). It would be possible to directly mount the system on a vehicle that would hydraulically drive the electrodes. The depth of the survey is controlled by the spacing of the sending and receiving units. EM sensors can provide station measurements to a depth of 200 ft. Continuous data may be acquired to a depth of 50 ft. Although EM provides excellent lateral resolution (profiling), it provides only limited vertical resolution (sounding). Buried metal

site conditions, detect soil conditions such as hardpan. The MAG's response is a function of an object's depth and mass. MAGs can provide station or continuous profile measurements and may be hand or vehicle mounted. MAGs can be used to detect a single drum to a depth of 18 ft and can be used to detect large masses of drums to depths of 18 to 80 ft. Magnetometers are the sensor of choice for many UXO sites. Portable magnetometers have been shown to be a particularly effective tool for locating, characterizing and identifying buried exploded ordnance (EOs) and UXOs that are composed of ferrous materials.

For example, currently millions of acres of Department of Defense sites are contaminated with ordnance as a result of troop-training and weapons-testing activities. Magnetometer methods, when coupled with GPS technologies and the increasing knowledge database of characteristic EO and UXO signatures, provides a method for swift preliminary characterization and mapping of areas of concern. However, the latter by itself cannot be taken as a magic black box but must be supported by limited ground-truthing exercises that verify signatures and interpretations, and facilitates screening out of site-specific influences that could generate false positive (be it natural soil conditions such as magnetic within-the-soil/ rock matrix or other man-made background noise, such as railroad tracks, pipes, fences, vehicles, buildings, etc.). Discrimination ability does not depend on the relative depths at which objects are located *vis-à-vis* interference (natural or man-made). Ordnance objects can be characterized beneath conductive overburden that might otherwise shield the ordnance objects from detection. Noise cancellation techniques may vary depending on-site conditions and include low pass and match filtering, averaging over many pulses, and using a remote set of sensor coils to measure ambient noise from the total target sensor signal. This sequence can be critical to lowering the false positive with subsequent impact on clean up efficiency and safety.

In summary, although some consider magnetometer data overly complicated and prefer EM sensors, magnetometers are appropriate (if not preferable) when nonferrous targets are not a concern and surface debris conditions are not significant.

Gradiometers are magnetic field sensors configured to measure the change (or gradient) of the magnetic field at a particular location. Similar to magnetometers, these devices are useful for detection of ferrous targets such as UXO and metal utility lines. The gradient measurement has an advantage over total field magnetometers by mitigating the effects of metal structures located in the vicinity of buried targets. Radiometers are also useful in congested areas where interference from

76   *Macroengineering: An environmental restoration management process*

5.4.1 Navigational methods

To fully exploit the data from geophysical sensors, the data must be collected with an accurate and reliable navigation technology. A map of subsurface targets and contamination is useless without accurate positional
accuracy. Advanced differential GPS technologies provide the solution to
this location problem with robust, half-foot, real-time accuracy. Currently,
the most appropriate differential GPS available for field-mapping applications is the Trimble Pathfinder XRS. This system offers full integration with
geophysical sensors, real-time differential solutions based on either satel-
lite-provided differential corrections or base-station-provided differential
corrections, and the breakthrough multipath-rejection capability that
enables GPS positioning in tree-covered sites or near buildings. However,
several site-specific issues must be resolved before GPS can be used reliably
on the site. These include definition of the site-specific coordinate system
on all navigation equipment, establishment of a differential GPS base station, establishment of methods to utilize real-time differential corrections
over large sites, and complete testing (in the field) of these site-specific
modifications.

In addition to mapping geophysical data, GPS may be used for many
different characterization tasks, including:

- *Feature identification:* One of the most powerful ways to improve
  effectiveness of geophysical mapping of complex, poorly documented sites is to simply walk over the site and make observations. During
  this process, GPS plays a key role in position-stamping debris piles,
  unidentified features, soil changes, vegetation, burn areas, craters,
  stained areas, etc.
- *Digital photography:* If a picture is worth a thousand words, then a
  digital photograph position-stamped with DPGS must speak a million words. There is simply no better way than digital photography
  and GPS to document site conditions.
- *Grid-corner locations:* Differential GPS provides a simple, effective,
  and reliable method for defining survey grid corners. Using L1 frequency GPS in carrier phase mode provides 10 to 20 cm accuracy,
  sufficient for this task.
- *Target relocation:* GPS provides an exceptional tool for target/object
  relocation. The researcher can streamline the process of taking targets
  from the GIS by loading them into the GPS handheld unit. The
  waypoint mode facilitates quick and reliable relocation.

great deal of excitement has surfaced within the GPS community in the past few years with the advent of multipath rejection GPS technology. With the new hardware and software advances, it is now possible to survey in lightly and moderately wooded areas, under defoliated conditions, with 1 to 3 foot accuracy.

However, a word of caution: not all GPS integration techniques are equivalent. Simply logging GPS data during a survey and merging the data later can be extremely problematic. Inconsistent and unreliable time delays are common problems plaguing these efforts. Timing errors during data acquisition translate into spatial errors. These issues can be resolved through rigorously documented clock calibrations performed on each instrument at least twice daily. Anything less yields inaccurate target locations, increases false alarms, and the real possibility of leaving UXO, contaminant, or hazardous material in the ground.

5.4.2 *Geophysical data analysis*

Success in geophysical mapping and site characterization is dependent on high-quality data collected from the most appropriate sensors with sufficient navigational accuracy. However, another key to success is the application of sophisticated data analysis methods by experienced scientists (e.g., geophysicists). For example, two types of tools are being developed to determine UXO target parameters such as depth and weight. First, a signature database developed from previously excavated UXO is being used to fingerprint identified anomalies. This database is the underlying basis of the ordnance and explosives knowledge base (OE-KB) technology developed under the Army Corp of Engineer's Huntsville office sponsorship. The OE-KB has been used in over 20 live sites, as well as in over a dozen test facilities and demonstration. Comparable database development could be enacted for hazardous waste management areas. Second, an investigator might employ model-matching algorithms estimating size, depth, and orientation of targets based on best-fitting numerical model. Both methods have proven highly accurate when data quality is good, and can determine target depths within 0.5 ft.

Standard data processing includes data leveling, statistical data assessment, grid generation, and noncustomized data filtering to accentuate target signatures. All grids are loaded into the GIS for review and target identification. All detected targets are located and characterized to estimate target size/depth.

Advanced geophysical data processing starts after the execution of standard data processing. All data sets are reprocessed with adaptive filters

more time consuming than automated threshold methods as each target requires specific examination and review. Additionally, filter parameters often require adjustments in an interactive search and investigation of suspect signals. After all targets are detected, the data are processed to extract a geophysical signature from the data grid (typically a 20 ft × 20 ft box surrounding the target). All signatures are modeled with both analytical (model matching) and empirical (database lookup) methods. All results are presented to the analyst via the GIS for review and final recommendations for target parameters.

5.4.3 *Geophysical quality assurance issues*

The researcher should recognize that most sensitive detectors do not exhibit good discrimination or the ability to distinguish between UXO and non-UXO objects without the benefit of a strong signature database strengthened by a strong field-truthing program. The latter provides invaluable benefit to a successful data analysis quality assurance program. Geologic factors, moisture, and other site-specific phenomena can contribute to high false positives, as depicted in Figure 5.5. Just as a limited test fill can provide invaluable benefit to controlling a large-scale geotechnical engineering program by identifying the unique quirks and characteristics of geotechnical source material and site foundations, so a limited geophysical signature/site field-truthing protocol can enhance the understanding and interpretation of signature data.

A critical QA management tool is a well-designed and maintained site computer database to track the type and processing signature of cultural resources, UXO, species, environmental conditions, etc. From a quality engineering standpoint, the detection probabilities are probably in the maximum

Equipment	Environment	Measurement
• Calibration	• Geologic features	• Sensitivity
• Battery condition	• Weather	• Discrimination
• Type of detector	• Terrain	• Location accuracy
Meetings	**People**	**Materials**
• Sweep line speed	• Training	• UXO condition
• Sweep line spacing	• Fatigue	• UXO class

Chapter 5: Site characterization 79

range of 85% to 95%. However, the clearance goal in UXO work is 100%. This situation does not meet the traditional definition of a capable measurement system. As a consequence, an ongoing monitoring effort is required to improve the process over time. To that end, a quality assurance project plan (QAPP) is also needed to document the type and quality of the data used for decision making. The QAPP should specify how the data should be collected, assessed, analyzed, and be reported. It should include all aspects of the project that can affect data quality, including data precision, integrity, traceability, as well as calibrations, self-audits, and corrective actions. Ongoing assessment of the process must take place throughout the implementation of remediation activities.

DGM quality control is also performed to ensure that: (1) valid operation of all deployed equipment, (2) consistent, excellent data quality, and (3) reliable, repeatable, objective, and defensible data analysis results. This process can be broken down into numerous steps and documented SOPs. The basic operations involved in QC process can be broken into activities associated with the actual collection of data in the field and office activities such as data download, processing, production of dig lists, and deliverables. Example lists are presented in Figure 5.6, Figure 5.7, and Figure 5.8.

Crew Deployment Log: This log defines the location of each geophysical survey crew on a daily basis. The log tracks crew members, equipment, and expected area to be surveyed. Attached to this daily log are maps of the areas to be surveyed containing the coordinates of benchmarks in the areas as well as the coordinate of each quadrant corner.

•**Field Log:** This log is filled out by each crew chief and details all activities of the survey. This is a daily log and contains observations about crew performance, sensor performance, site conditions, soil conditions, and weather changes.

•**Instrument Calibration Log:** This log documents the daily calibration of each field instrument. Daily calibration procedures are executed for each geophysical and navigational instrument. The sensor system is brought to a calibration area before each survey day starts and the background magnetic field and the magnetic field signal from a reference target is measured and recorded.

•**Data Control Log:** Kept in the data control center, this log tracks all data flowing in from the field and out to the SC&A office. Data include all geophysical field data, calibration data (via Calibration Logs), all field notes from Field Logs, and all GPS quadrant coordinate data. This log tracks the GIS system electronically, with hard copy prints made daily.

•**Data Processing Log:** All magnetometer data from the field are run through a standard data-processing procedure. This procedure is the same for all data and is tracked with the

80   *Macroengineering: An environmental restoration management process*

• Instrument serial numbers are recorded in field logs.

• Personnel are checked for metallic objects prior to survey commencement.

• Wiring is secured to the transport structure to minimize noise directly from the instrument.

• Azimuthal measurements are made to determine any dependence of the measured signal on azimuth and corrections are applied to measurements obtained along different azimuths, as necessary.

• Navigation instrumentation is calibrated over a known monument.

• Instrument calibrations are performed, recorded, and logged morning and evening over a known source to ensure that instrument functionality is maintained within the required specifications of repeatability.

• Individual measurements are compared to the locally obtained statistical baseline information to determine the normal operating range and deviations that constitute failure.

• System timing delays are determined from the calibration data and corrected to ensure accurate positioning.

• Tick wheel operation and/or fiducial marks are used as a primary or backup method of positioning when GPS or acoustic methods cannot be applied or fail in the field.

• Instrument transport structures are maintained level to ensure consistent positioning and data.

• During grid operations, the first and last lines are repeated in opposite directions to insure instrument and data quality.

• GPS features are recorded for each individual grid and meander path to serve as a backup record independent of the field log and field maps.

• Field geophysicists and instrument operators continuously check instrument readouts and audio alerts to ensure proper operation.

Figure 5.7 Field operation QC procedures.

9202_C005.fm Page 81 Thursday, January 12, 2006 11:04 AM

Chapter 5: Site characterization   *81*

- •Data are downloaded to PCs with data quality assessment and filename recorded in the processing log for each file.

- •Visual QC by a data processor follows conversion of the files to the adf and grid formats.

- •The effects of leveling and noise reduction operations are evaluated by a data processor following each step and are periodically reviewed by a senior geophysicist.

- •Visual QC of automatic and/or manual target picks is accomplished by a data processor and periodically reviewed by a senior geophysicist by overlaying them on the geophysical data.

- •A systematic and consistent numbering scheme is utilized to simplify QC of target numbering.

- •A senior geophysicist and at least one other analyst reviews all deliverables to identify any inconsistencies, errors, or omissions.

- •Separate records of deliverables are maintained to permit historical review and tracking of any changes in target lists.

Figure 5.8 Data download, processing, and deliverables QC.

Bibliography

1. U.S. EPA, Field Screening Methods Catalog, User's Guide, EPA/540/2-88/005, Office of Emergency and Remedial Response, Washington, D.C., 1988.
2. U.S. EPA, Field Screening Methods for Hazardous Waste Site Investigations, First International Symposium, 1988.
3. Technos, Inc., Contemporary Site Investigation Techniques, Geophysical Methods, Vendor Information, 1985.
4. U.S. DOE, Office of Environmental Management, Office of Technology Development, Characterization, Monitoring, and Sensor Technology Integrated Program (CMST-IP), Technology Summary, Washington, D.C., February 1994.
5. U.S. EPA, Subsurface Characterization and Monitoring Techniques; A Desk Reference Guide: Rept. EPA/625/R-93/003, by Eastern Research Group, Lexington, MA, May: v.@: The Vadose Zone, Field Screening and Analytical Methods; Appendices C and D: Rept. 003B, 405 p, PB94-131497, 1993.
6. Foley, J.E., Environmental characterization with magnetic and Surface Tower Ordnance Location System (STOLS); *Proceedings of the IEEE*, Vol. 82, No. 12, December, 1994.
7. Kadzie, D., Lantzer, N., Wernsman, R., Sieber, E., and Bond, T., Aftermath of

chapter 6

Discussion on mobile laboratory requirements

For characterization and confirmation testing during the actual restoration phase, Macroengineering favors having an on-site mobile laboratory setup to expedite the analytical work with a percentage of split samples being sent off-site to confirm the accuracy of the mobile laboratory data.

The use of a mobile screening laboratory presents several advantages over off-site analysis, including both cost and time savings. First, costs can be decreased in some cases by a factor of ten vs. the use of off-site contract laboratory programs (CLPs) for analysis. Secondly, utilizing the mobile laboratory will afford time savings in all phases of the analysis process, including decreases in holding times, transportation to off-site facilities, coordination of deliverables, sampling analysis and scheduling and, if necessary, mixed waste and Health Physics Technician (HPT) coordination. These time savings can be significant when considered in view of the magnitude of a project such as a typical macroengineering remediation site.

Mobile laboratories currently have the capability to process both organic and inorganic wastes. Waste streams that can be analyzed include the following:

- Organic:
  Volatiles, high vapor pressure, small number of carbons, CCl_4, hexone, and benzene
  Semivolatiles, low vapor pressure, larger number of carbons, phenols, PCBs (aroclor), DDT, and nitrosamines
- Inorganic:
  Metals
  Cations
  Anions

This section lays out the functional design criteria and instrument specifications of the mobile screening laboratory.

6.1   Physical structure requirements

The mobile laboratory structure should meet the following specific requirements.

- The laboratory should be wheel mounted and, when towed by a vehicle, should be able to negotiate unimproved dirt or gravel roads. Trailers provided by the vendor should be licensed for use on state highways.
- The laboratory should be equipped (either internally or externally) to provide the necessary power while the mobile laboratory is in remote locations. One generator is needed to provide high-quality power for all of the analytical instruments and required computer hardware. The other generator should be used to operate the lights, heating and air conditioning systems, exhaust hoods and fans, and the remaining environmental power needs. Because the generators required will be quite large, a separate trailer for these may be optional pending the vendor design of the mobile screening laboratory. The laboratory should also have the capability to accept power from conventional sources when available.
- A reliable source of water will be required for analysis. A central reservoir should be incorporated within the mobile laboratory facility to house distilled water. The reservoir will supply distilled water, which will be fed through an anion and cation exchange resin deionizing system, and then through a high-intensity ultraviolet light/peroxide system to produce analytical-grade carbon-free deionized water. The purity of the analytical-grade carbon-free deionized water should meet or exceed Type I, as specified in ASTM D 1193, having a maximum conductivity of 20 & L mhos cm^1 at 250°C. All water sources must be housed in such a way that the water temperature will be the same as the interior ambient temperature of the mobile screening laboratory.
- The laboratory should be anchorable to withstand winds of speeds as high as 100 mi/h and will be sealed to prevent excessive dust entry.
- The laboratory structure should meet the applicable requirements for fire protection.
- The ventilation system and fume hoods should meet the requirements of the American Conference of Governmental Industrial Hygienists (ACGIH) ventilation manual.

- If applicable, given site characteristics, appropriate radiation safety design requirements must be met such as DOE Order 6430.1A, Section 1325,
  laboratory facilities (including hot laboratories). For example, the
  laboratory could be restricted to examining prescreened samples not
  to exceed any of the following radiation levels: 5 mrem/h, 50,000 dpm
  beta, or 5000 dpm alpha.

6.2   Specific requirements related to instruments and analytical capabilities

Specifications for the mobile laboratory should be developed based on EPA
requirements and a review of other studies relating to mobile laboratories.
The equipment and analytical capabilities required in the laboratory are
specified in the following.

6.2.1   EPA requirements

There are five EPA levels of analytical requirements for determining the
extent of environmental pollution (Appendix B, Data Quality Objectives for
Remedial Response Activities, EPA/540/G-87/003).

- *Level V:* This is the highest level of analysis by nonstandard methods.
  Analyses are performed in an off-site analytical laboratory. Method
  development or modification may be required for specific constituents or detection limits. Included in this level are CLP special analytical services (SASs).
- *Level IV:* This level of analysis is CLP routine analytical services
  (RAS). This analysis requires that data could be used for litigation
  and regulatory enforcement purposes. This type of analysis is extremely labor- and instrument-intensive and is characterized by a
  high degree of precision and accuracy.
- *Level III:* This level of analysis is performed in an off-site analytical
  laboratory and may or may not use the CLP protocol. The validation
  or documentation of procedures required of CLP Level III is optional.
  This level is used in support of engineering studies using standard
  EPA-approved procedures.
- *Level II:* This level of analysis is characterized by the use of portable
  instruments that can be used on-site, or in mobile laboratories
  supporting a field investigation. There is a wide range in the quality

Both Level II and Level I are conducted in a timely fashion and are used to select those samples to be sent for the highest level of analysis in the CLP. Instruments will be selected to provide Level II and Level III soil and water analysis of the known and suspected site contaminates of the given site.

6.2.2   *Literature review of mobile screening laboratories*

The rationale for the choice and configuration of the mobile screening laboratory instrumentation should originate from current and future environmental restoration needs of the site and from the following mobile laboratory literature review.

EPA publication 600/X-84-170, *Survey of Mobile Laboratory Capabilities and Configurations*, encompasses several facets of mobile laboratory design and operation. Design characteristics of several mobile units are presented, and include specifications for ventilation, power, heating/cooling, safety equipment, and vehicle suspension. Sophisticated GC/MS systems are the primary instruments of analysis in the presented mobile laboratory configurations.

EPA publication 540/2-88-005, *Field Screening Methods Catalog: User's Guide*, is a compilation of the methods that have been identified as being used in EPA regions for field screening. Given a specific site characteristic, the user is able to identify a field screening method with appropriate instrumentation.

EPA Regional Guidance, for example, *FIT Field Analytical Support Program Cost Analysis*, EPA Contract 68-01-7374, is a report prepared by Ecology and Environment, Inc. This report documents the cost effectiveness of Field Analytical Support Program (FASP) procedures used by EPA Region X in support of Superfund preremedial activities. This report contains FASP expenditures from 1984 to August 1988, estimates FASP cost per sample group, processing rates, and cost effectiveness of FASP vs. CLP analysis.

Proceedings of the Fifth Annual Waste Testing and Quality Assurance Symposium, a joint presentation by the American Chemical Society and the EPA, highlight the areas of quality assurance (QA) and analytical method development and evaluation. Emphasized is the methodology and practices that are being developed or applied to implementing the Resource Conservation and Recovery Act (RCRA) and the Comprehensive Environmental Response, Compensation and Liability Act (CERCLA) hazardous waste management programs.

Field Investigation Team (FIT) Screening Methods and Mobile Laboratories Complementary to Contract Laboratory Program (CLP) is a joint publication of the NUS Corporation and Ecology and Environment, Inc. This document

applications. Three mobile laboratory configurations are presented that offer Level I, Level II, and Level III analysis capability. A lease-vs.-buy cost analysis is also presented.

6.3   Analysis process

A critical element in preparing a mobile laboratory specification is the analysis process required by the site. Figure 6.1 shows an example of the flow of the mobile laboratory waste analysis process for a mixed-waste environmental restoration site and as illustrative of the thought process that must be undertaken. In the example, the mobile laboratory utilizes the supercritical fluid extraction (SFE) process for volatile, semivolatile, and nonvolatile organic analysis in place of the Soxhlet extraction procedure. SFE is less costly, faster, and more environmentally conscious than the alternative procedures such as Soxhlet. The ability to perform on-line or off-line analysis with minimal space requirements would be another advantage of SFE.

Contaminants in the soil originating from man's deposition of industrial waste can be classified into various categories. Each category can then be analyzed for quantification by techniques and instrumentation specific for the category of interest. Categories that have been the focus of mobile screening laboratories include organics (volatiles, semivolatiles, and polycyclic aromatic hydrocarbons (PAH)), inorganics (metals), radionuclides, pesticides, and PCBs.

Tables should be developed in the site sampling and analytical plans that clearly identify for each analyst the on-site analytical method, required precision, required accuracy and the required PQL.

6.3.1   Sensitivity

Sensitivity is achieved in the laboratory using instrument detection limits (IDLs), method detection limits (MDLs), and practical quantification limits (PQLs). These limits are published with U.S. Environmental Protection Agency (USEPA) methods. They are based on a reagent water matrix and do not account for specific sample matrices. IDLs are generally not required under SW-846 methods; however, they are required for USEPA Contract Laboratory Procedure protocols and when performing SW-846 Method 6020. The IDL samples estimate the instrument's detection limit under ideal conditions. The IDL samples are introduced at a later stage of the analytical process where instrument sensitivity can be directly measured. MDLs estimate the detection limits by introducing a known concentration matrix to

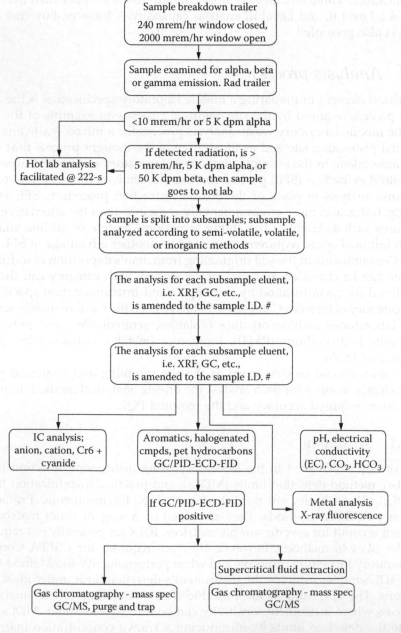

Figure 6.1 Example of mobile laboratory specification flow chart per macroengineer-

6.3.2   Organic contaminant analysis

One of the key components of a mobile analytical laboratory is a gas chromatograph system. This system has the ability to perform rapid identification of volatile and semivolatile organic contaminants.

In gas chromatography (GC), a volatile liquid or gaseous solute is carried by a gaseous mobile phase. The stationary phase is usually a relatively nonvolatile liquid coated on a solid support (the interior of the GC column). The volatile liquid sample is injected through a rubber septum into a hot glass or metal-lined injector port, which vaporizes the sample. The sample is then swept into a column by an inert gas (usually He or N_2), which serves as the mobile phase. In passing through the column containing the stationary phase, the solutes are separated from each other as each solute equilibrates with the stationary phase. The solute which has the greatest affinity for the stationary phase will move through the column more slowly. The gas stream flows through a detector, which sends a signal to a recorder as solutes emerge from the column. Identification of the solute (or eluent) constituents requires the use of detectors sensitive to the chemical grouping of the constituents, i.e., aromatic hydrocarbons, halogenated hydrocarbons, etc., which are discussed in a screening perspective in the following section.

6.3.2.1   Screening with GC/FID, GC/PID, GC/ECD, and GC/ELCD

Several types of GC detection systems are useful in the screening of contaminated media where detected contamination is further quantified by a gas chromatograph/mass spectrometry (GC/MS) system.

The gas chromograph/flame ionization detector (GC/FID) can be used for qualitative determination of hydrocarbon contamination. This detection is specific for compounds containing carbon and hydrogen, but not the functional groups attached to the carbon chain. Thus, two entirely different compounds that are the same in the number of carbons but different in the functional groups or carbon chain configuration would conceivably be identified as the same compound. Some of these carbon chain anomalies, the aromatic hydrocarbons in particular, can be detected by use of a gas chromatograph/photoionization detector (GC/PID). Chlorinated hydrocarbons such as carbon tetrachloride and chloroform, unsaturated hydrocarbons (benzene, ethylene, etc.), and chlorinated hydrocarbons that have a double bond, such as trichloroethylene or perchloroethylene, are more easily detected by this type of detection scheme.

A gas chromatograph/electron capture detector (GC/ECD) has special utility in that it can be set up as a screening device for the halogenated compounds or can be made to quantitatively determine concentrations of

90   *Macroengineering: An environmental restoration management process*

ELCD detection of halogenated compounds a better choice for samples in which the kind and extent of halogenated compounds are not known.

Amending any of these detection schemes by the use of a wide-bore capillary column (rather than a packed column) enables a large flow through of sample without having to dilute before injection. This provides a quali- tative determination of the amount of organic contamination and is advan- tageous in determining the dilution factor of the sample before GC/MS analysis. Important to field screening is the ease of operation, instrument down-time, turn-around time, and real-time analysis. A gas chromatograph configuration with multiple in-series detectors accommodates quick turn-around time and real-time analysis. Using one sample injection in place of three reduces the number of GCs needed, expedites the analysis process, and requires less sample for total analysis. As an example, a PID, ECD, and FID could be configured in series of one GC, requiring only one injection for qualitative determination of aromatic, halogenated, and petroleum volatile compounds. Few portable GC units on the market have this multiple detector utility.

6.3.2.2   *Methodology and detection limits for GC/PID, GC/ECD, and GC/FID screening*

There is not a prescribed methodology for the series configuration of GC screening using multiple detectors, but a composite method derived from the methods used for GC/PID, GC/ECD, and GC/FID analyses should be employed. Because this GC configuration is typically used as a preliminary screening before GC/MS analysis, the detection limits need not be as sensi- tive as methods described for levels of analysis greater than Level III. Because each detector registers for a respective group of hydrocarbons, three separate chromatograms will result. Consequently, detection limits should be pre- scribed for each detection scheme.

Methods FM-6, FM-8, and FM-10 from EPA publication 540/2-88-005 *Field Screening Methods Catalog: User's Guide*, should be combined and used as the basis for this analysis method. The MDLs for those compounds detected by the PID, ECD, and FID should be no less than 100 Fg per kilogram of soil and 100 Fg per liter of water. The analysts should consult these methods and SW-846 Method 8015 (nonhalogenated volatile organics using FID), Method 8010 (halogenated volatile organics using a halogen-specific detector), and Method 8020 (aromatic volatile organics using PID) for cali- bration, standardization, maintenance, and waste disposal.

6.3.3   *Volatiles vs. semivolatiles*

compounds that are gaseous at 25°C and 1 atm (CH_4 through C_4H_{10}) to liquids (C_5H_{12} through $C_{16}H_{34}$) to low-melting paraffin-like solids, the boiling point increases. Recalling the principle in GC in which the volatility of the compound is increased to induce the chromatographic separation, it becomes apparent that some organic compounds require a higher temperature to induce volatility. Those organic compounds requiring higher temperatures to induce volatility are referred to as semivolatile organics. The organic compounds requiring less temperature input to induce volatility are referred to as volatile organics. This characteristic of organics dictates two different analysis procedures.

The analysis of volatile organic compounds is commonly implemented using the purge and trap extraction technique, followed by GC/MS detection. It is the required technique for a number of EPA methods for analysis of drinking water, source and wastewater, soils, and hazardous waste. In this method, samples contained in gas-tight glass vessels are purged with an inert gas, causing volatile compounds to be swept out of the sample and into the vapor phase. Organic compounds are then trapped on an absorbent, which allows the purge gas and any water vapor present to pass through. The volatile compound can be efficiently collected from a relatively large sample, producing a concentration factor that is typically 500- to 1000-fold greater than the original. After collection, the adsorbent is heated (thermally or in a microwave oven) to release the sample and then backflushed using the GC carrier gas. This sweeps the sample directly into the GC column for separation and detection by normal GC procedures.

Traditional analysis of semivolatile organic compounds involves time-consuming extraction processes followed by some sort of GC detection. Extraction of semivolatiles from liquid and soil/sediment using CLP protocol (EPA Method 625) or SW-846 Method 3540 requires hot water baths, heating plates, sanitization equipment, specially designed glassware, and large volumes of the extraction fluid, methylene chloride.

The space and equipment requirements are extensive, and the extraction procedure can take up to 24 hr, the extracted sample will be concentrated prior to analysis, and large volumes of spent methylene chloride extraction waste accumulate. These characteristics of the semivolatile organic exit-action methods reduce the applicability to a quick turn-around analysis scenario, as would be desired in a mobile laboratory screening situation. An extraction method that is more rapid and provides cost-effective assessments of semivolatile organic contamination would be preferred.

SFE is a rapid and cost-effective extraction technique for the analysis of semivolatile organics. The SFE technique provides a viable alternative with distinct advantages over the liquid extraction methods.

92   *Macroengineering: An environmental restoration management process*

- Because SFE does not utilize the large volumes of the methylene chloride extraction solvent, there is minimal solvent waste.
- Common fluids used in the SFE process, i.e., CO_2 and N_2O, have low critical temperatures, and allow extractions under thermally mild conditions, thereby protecting thermally labile components.
- Because supercritical fluids undergo expansive cooling upon decompression (part of the extraction process), even volatile organic compounds can be quantitatively and efficiently collected into solvents (off-line) which can be analyzed by GC methodology.
- Common fluids, used in the SFE process, i.e., CO_2, N_2O, and SF_6, are gases at room temperature, which further simplifies off-line collection.
- The extracted compounds from the SFE process can be directly injected into the GC for analysis (online analysis). No sampling handling is required between extraction and GC analysis. This increases the quality and reproducibility of the data.
- The extraction time is usually less than 30 min. When compared to the actual GC/detector analysis time, the extraction process is no longer the rate-limiting factor.

The extracted nonvolatile and semivolatile compounds from the SFE process will be directly injected into the GC/MS system for analysis.

6.3.3.1 *Methodology and detection limits for volatile and semivolatile organics analysis*

The EPA SW-846 Methods 8240 and 8260 should be used for GC/MS analysis to quantify most volatile organics that have boiling points below 200°C. Included in this category are low-molecular-weight halogenated hydrocarbons, aromatics, ketones, nitriles, acetates, acrylates, ethers, and sulfides. The actual detection limits of volatile analysis for a particular brand of GC/MS are difficult to establish. The vast number of compounds analyzed, column setup, carrier gas, and purge and trap configuration all contribute significantly to the establishment of detection limits. Most major GC/MS companies (HP, Finnigan, and VG Instruments) comply with the detection limits using EPA 40 CFR 136 Method 624. For most of the detectable volatile organics, the method detection limits should be below 10 Fg L^1. The SW-846 Methods 8240 and 8260 stipulate instrument setup specifications for detection limits of volatiles concurrent with 40 CFR 136 Method 624. Therefore, the observed detection limits for volatile organic GC/MS analysis should be according to these methods.

For GC/MS analysis of semivolatiles, EPA SW-846 Method 8250 and

detection limits for volatiles also apply to the GC/MS used for semivolatile organic compound analysis (the purge and trap configuration is not included in semivolatile analysis). Most major GC/MS companies (HP, Finnigan, and VG Instruments) comply with these detection limits using EPA 40 CFR 136 Method 625. For most of the detectable semivolatile organics, the method detection limits should be below 10 Fg L¹ Method 8250 and Method 8270 from SW-846 stipulate instrument setup specifications for detection limits of semivolatiles concurrent with 40 CFR 136 Method 625. Therefore, the observed detection limits for semivolatile organic GC/MS analysis will be according to these methods.

6.3.4   Metal contaminant analysis

Inorganic pollutants do not exemplify the vapor pressures needed for GC determination. They can be readily determined by other instrumentation, most notably atomic absorption/emission spectrophotometry (AA/AE), inductively coupled argon plasma spectrophotometry (ICP or ICAP), and energy-dispersive x-ray fluorescence spectrophotometry (XRF or EDXRF). All these methods have been used in some capacity for preliminary screening of metals from hazardous waste sites.

Simultaneous multielement analysis, rapid turn-around time, and ease of sample preparation for analysis are important to field screening. The use of AA/AE analysis does not allow for multielement analysis and requires an acid extraction sample preparation. The use of ICP facilitates multielement analysis, but requires the destruction of the sample (acid extraction). It is also very sensitive to external vibrations; when used in a field screening situation, precautionary measures should be taken to prevent physical shock to the instrument. Until recently, the detectors of the XRF units were cooled with liquid nitrogen, making a difficult and cumbersome configuration in a mobile screening laboratory. This problem is circumvented in modern units by the use of thermoelectrically cooled silicon detectors.

The great advantage of XRF is the simultaneous detection of 46 metals (atomic number >11), 18 of which have EPA priority on the hazardous substances list. In addition, XRF is well suited for screening analysis because it: (1) requires minimal sample preparation; the sample is not destroyed, and it can be stored or used for additional analyses; (2) provides rapid turn-around time; (3) has dynamic range that corresponds to typical soil contamination (ppm to 100%); (4) can be used at normal atmospheric pressure for analysis of solid, liquid, or gas samples; and (5) furnishes an accurate quantitative analysis with results comparable to those obtained using CLP

94   *Macroengineering: An environmental restoration management process*

Table 6.1 Tracor Spectrace 6000 Detection Limits
for 18 Target Elements

Metal	Spectrace 6000	
	Soil $\mu g\ g^1$	Water $\mu g\ L^1$
Antimony (Sb)	20	250
Arsenic (As)	20	200
Barium (Ba)	20	250
Cadmium (Cd)	20	250
Chromium (Cr)	20	600
Cobalt (Co)	20	600
Copper (Cu)	20	200
Iron (Fe)	20	400
Lead (Pb)	20	100
Manganese (Mn)	20	600
Mercury (Hg)	20	100
Nickel (Ni)	20	200
Selenium (Se)	20	150
Silver (Ag)	20	200
Thallium (TAW)	20	150
Tin (Sn)	20	250
Vanadium (V)	20	600
Zinc (Zn)	20	200

6.3.3.2 *Methodology and detection limits for XRF metal analysis*

The selected method for XRF metal analysis in the mobile screening laboratory is FM-2, listed in EPA/540/2-88/005, *Field Screening Methods Catalog:
User's Guide*. Table 6.1 lists the specified detection limits using the Tracor
Spectrace 6000 XRF unit for selected priority metals in a soil matrix and in
a water matrix. Fundamental limitations of the XRF technique prevent the
analysis of boron and beryllium. Aluminum is usually not analyzed because
of low instrument sensitivity and ubiquitous nature in the soil environment.
Iron is usually not analyzed for the same reason and is seldom an element
of critical concern.

6.3.5 *Cation and anion analysis*

Not all of these compounds can be analyzed by using the GC/MS and XRF
configuration. In particular are the ions NO3, NO2, Cl, F, Br, SO_4^2, HPO_4^2,
Cr^{6+}, and CN. Measurements of anions are important from a remediation
aspect in that anions act as ligands in potential metal ligand complexes.

site may feed into drinking water municipalities, these ions may be of concern for the remediation of the site. Method 300.0 of the EPA document, *Methods for Chemical Analysis of Water and Wastes,* designates ion chromatography as the analytical principle applicable to the simultaneous determination of NO_3, NO_2, Cl, F, SO_4^2, and HPO_4^2 in drinking water, surface water, and mixed domestic and industrial wastewater. This method dictates a holding time of 48 h at 4°C for NO_3 analysis. This makes transportation of sample for off-site analysis quite difficult. An analytical facility on-site would eliminate the need for shipment of samples to an off-site laboratory for NO_3 analysis. The mobile screening laboratory could house ion chromatography equipment that could easily facilitate the determination of NO_3, NO_2, Cl, F, Br, SO_4^2, HPO_4^2, and some of the soluble cations, such as Na^+, NH_4^+, K^+, Mg^{2+}, and Ca^{2+}, determined for drinking water standards and for remedial investigation.

The selected methods for the ion chromatographic analysis of anions and cations using the Dionex DX-100 ion chromatography system are 300.0 (anions), 300.7 (cations), and 353.2 (nitrogen, NO_3/NO_2). Detection limits for NO_3, NO_2, Cl, F, Br, SO_4^2, HPO_4^2, Na^+, NH_4^+, K^+, Mg_4^+, and Ca^{2+} are method specific and should be below 20 Fg ml^{-1}. Determination of Cr^{6+} should be accomplished using the methodology from Dionex Corporation, Method TN26. Determination of CN should be accomplished using methodology from Dionex Corporation, application update # 107. Methods for the ion chromatography determination of Cr^{6+} and CN, developed by Dionex, are under review by the EPA. These detection limits for Cr_{6+} and CN are method specific and shall both be below 10 Fg L^1.

Table 6.1 lists the 18 target elements regulated by the NPDES, the Resource Conservation and Recovery Act (RCRA), and the CWA that will be screened in the mobile screening laboratory by x-ray fluorescence. The second column lists the required detection limits in soil and water using the Tracor Spectrace 6000 XRF unit.

6.3.5.1 Auxiliary equipment

For the GC methods, XRF methods, and cation and anion analysis, pH measurements shall be made. Electrical conductivity of solutions is necessary, and the concentrations of CO_2 and CO_3 will be known. Examples of instruments which provide these measurements are provided in Table 6.2 and should be included in the mobile screening laboratory.

6.3.6 Radiation sample analysis flow

The sample should be tagged for identification as soon as it is pulled from

Table 6.2 Auxiliary Instruments for pH, Conductivity, Weighing,
and Solution Makeup

Instrument
Balance, Mettler PM 200 electronic top-loading balance
Balance, Mettler AE 100 electronic analytical balance
pH meter, Accumet 915 pH/mV/temperature
pH electrode, Orion Ross series 810200
Conductivity meter, YSI model 34
YSI 3417 conductivity cells for YSI model 34
YSI temperature probe for YSI model 34
Carbon coulometer (CO_2 and CO_3 analysis)
Software package for collection, calculation, printing, and storage of the analytical data from the carbon coulometer
Installation and equipment setup for the UIC system 140
General laboratory supplies, glassware, and chemicals

10-g sample would then be routed through another radiation analysis. If this second radiation analysis is >5 mrem/h or 50,000 dpm beta or 5,000 dpm alpha, the 10-g sample will then be routed to either 222-s or 325 labs for analysis. If it is <5 mrem/h or 50,000 dpm beta or 5,000 dpm alpha, then this 10-g sample will be routed to the mobile screening laboratory, where it will be split into subsamples and analyzed for inorganics, organics, and/or cation–anion analysis.

Before each subsample analysis, the subsample should be tagged into the computer sample-tracking system. Please note, aside from the obvious health precautions in the analysis of a sample above the 5 mrem/h or 50,000 dpm beta or 5,000 dpm alpha cutoff limits, the XRF analysis of could be complicated in the presence of radionuclides. The x-ray beam excites the element; energy is emitted by the excited element, and is then detected by the detector. The gamma radiation from hot samples will foul the detection system in the XRF, causing false readings. After completion of analysis, the results will be input into the computer system.

6.3.7 Complete mobile laboratory analytical configuration

An example of a complete instrument configuration for mobile screening laboratory is presented in Table 6.3. Included are the instrument specifications for the organic, inorganic, cation, and anion analysis.

Chapter 6:   Discussion on mobile laboratory requirements   97

Table 6.3 Complete Instrument Configuration for the Environmental Division
Mobile Screening Laboratory

Volatile organic analysis

GC/MS, HP 5890A GC, HP 5970B MSD, MS
Chemstation QS/20, Target Environmental
CLP software, (Combined Part # G1021A)
NIST MS compound library
OI purge and trap, part # 177338
OI low dead volume interface, part # 176900
OI external carrier flow module
(ECM) part # 193128, cables

Semivolatile organic extraction

Suprex SFE system, SFE/50 Integrated SFE, Syringe pump, stepper motor, control unit
  with oven
Multivessel Package that includes four extraction vessels
Solvent modifier value
Autopump
SFE/GC transfer kit
Extraction vessels: 2 ml, 5 ml, and 8 ml
1-yr operation kit

Combined hydrocarbon, aromatic and halogenated organic screening

GC/PID-ECD-FID, SRI 8610 GC connected in series, wide-bore 0.53-mm capillary
  column
GC/PID-ECD-FID, SRI 8610 GC connected in series, wide-bore 0.53-mm capillary
  column (backup)
ELCD detector, Hall type, part # 8690-26, capability to be hooked in series with PID
Autosampler for soil and water GC screen, PRA-30 W/S, Dynatech Precision
  Sampling Corp.

Inorganic screening

XRF, Tracor Spectrace 6000 XRF
Vacuum pump
Power source
Microwave, for sample dry down before grinding
Soil grinder for improved accuracy, Spex Mixer Mill, Part # 8000-115, 8004A
Sample cups and window material Chemplex # 1430 cups and #437 polypropylene
  window material
Standards: CANMET SO-2, CANMET SO-3, CANMET SO-4, NIST 2704, NIST 1648,
  ERA 202, PACS-1

Table 6.3 Complete Instrument Configuration for the Environmental Division
Mobile Screening Laboratory (Continued)

Cation and anion screening
Dionex DX-100 ion chromatography system, thermal stabilizer, SPK, upgrade 2 columns, cable
Automated Sampler (ASM)
Polyvials and filter caps, 5 ml, 250 each, cassettes, 5 ml, box of 6
System control package, relay/TT1 cable advanced controller interface 3 function
Anion analysis: OmniPac PAX-100 column, OmniPac PAX-100 column guard column, AMMS-11, installation kit AMMS or CMMS, IonPac NG1 Guard column
Cation analysis: IonPac-CS10 column, IonPac-CS10 column guard column, suppressor CMMS-11 (cation chromatography), installation kit AMMS or CMMS

Auxiliary equipment
Balance, Mettler PM 200 electronic top-loading balance, Fisher catalog # 01-911-162
Balance, Mettler AE 100 electronic analytical balance, Fisher catalog # 01-909-375
pH meter, Accumet 915 microprocessor, pH/mV/temperature, Fisher catalog # 13-636-915
pH electrode, Orion Ross series 810200 Fisher catalog # 13-641-762
Conductivity meter, YSI model 34 Fisher catalog # 09-324-34
YSI 3417 conductivity cells for YSI model 34, Fisher catalog # 09-324-28
YSI temperature probe for YSI model 34, Fisher catalog # 09-324-42
Carbon coulometer (CO_2 and CO_3 analysis), UIC Inc., System 140
General laboratory supplies, glassware, and chemicals

6.4.1   Quality assurance

A QA engineer should be designated to be responsible for reviewing and advising on all aspects of QA/QC. Duties should include assisting the data requestor in specifying QA/QC procedures to be used in the mobile screening laboratory, designating the deliverable data format, making evaluations of QA/QC compliance, and submitting audit samples to assist in reviewing the QA/QC procedures. Upon encountering problems, the QA engineer should make recommendations to the appropriate level of site management to ensure corrective measures. The QA engineer should also be responsible for providing documentation to management (the data user) that will include:

- Routine assessment (surveillances and audits) of QA/QC measurement indicators
- Results of performance audits
- Significant QA/QC problems and recommended solutions

appropriate for analytical Level III methodology. Surrogate inorganic and organic compounds will not be required as part of the analytical QC requirements.

The basic and most important element of determining the level of quality is comparison against known standards. The ability to recover an accurately prepared standard is the fundamental assumption of laboratory QA. If done internally, the process is fraught with bias. To verify the performance of the mobile screening laboratory and the quality of the environmental analyses from an objective standpoint, the mobile laboratory should be assessed by the Proficiency Environmental Testing (PET) program such as conducted by the Analytical Products Group (APG), a subsidiary of Curtin Matheson Scientific, Inc. The PET program is the analysis of a specific set of standards supplied by APG and analyzed by the specific laboratory — in this case, the mobile screening laboratory. These same standards are also sent to other labs participating in the PET program.

Standards are issued twice monthly for the analysis of volatile and semivolatile organics, XRF metals, and ions analyzed by ion chromatography. APG then assesses the mobile analytical results and report to the QA officer and QA engineer the true values, average percent recovery, actual means and standard deviations of other participating laboratories, percent recovery, and the mobile screening laboratory deviation from the mean.

For every analytical batch, the following should be observed:

- Duplicate samples should be analyzed from every site, the frequency designated in the respective sampling plan.
- A reagent blank should be carried through for each of the analytical procedures.
- Each analytical batch should contain a check sample that will contain a representative subset of the analytes determined, the concentrations of which shall approach the quantification limit of the matrix of the check sample, which will also be used to determine the level of accuracy.
- All batches of adsorbents used in chromatographic analysis should be checked for analyte recovery by running the elution pattern with standards as a column check. This elution pattern should be optimized for maximum recovery of analytes and maximum rejection of contaminants.
- The analytical instrumentation should be tuned, aligned, and calibrated specific to the instrument and in accordance with the requirements specified in the analytical procedure utilizing the instrument.
- Additional QC for XRF is required because of the nature of the analysis. The duplicate sample runs will serve to determine both the

standard concentration falls out of the 35% relative deviation, the procedure is restandardized.

- Additional QC for organic analyses using GC/MS is required for the calibration and tuning. The calibration of each instrument should be verified at frequencies specified in the methods, as demonstrated by a standard curve. The tune of each GC/MS is checked with 4-bromofluorobenzene (BFB) for determinations of volatile organics and decafluorotriphenylphosphine (DFTPP) for the determinations of semivolatile organics. If the tune does not meet the specifications of any ion in the ion abundance criteria set forth in the method, the instrument will be retuned and rechecked and adjustments made before proceeding with sample analysis. This tune calibration will be checked daily or for each 12-h operating period.

- Additional QC for ion chromatography analysis is prescribed in EPA/600/4-79/020 *Methods for Chemical Analysis of Water and Wastes Methods* 300.0 (anions), 300.7 (cations), 353.2 (nitrogen, NO_3/NO_2), 218.1 (Cr^{6+}), and 335.3 (total CN), and should be executed for each of these analyses.

The detection limit and quantification limit of analytes shall follow the same calculating procedures stated in Chapter 1, Quality Control, SW-846; data reporting and quality control documentation should follow the same format as stated in SW-846, Chapter 1, Quality Control.

Bibliography

1. Moody, T.E., Instrument Specifications, Configuration and Functional Design Criteria of the Mobile Screening Laboratory.
2. U.S. EPA, Data Objectives for Remedial Response Activities, EPA/540/G-87 1003, OSWER Directive 9355.078. EPA Washington, D.C.
3. U.S. EPA, Survey of Mobile Laboratory Capabilities and Configurations, 600/x-84-170, Environmental Monitoring Systems Laboratory, Las Vegas, NV., 1987.
4. U.S. EPA, "Field Screening Methods Catalog: User's Guide," 540/2-88-005, 1988. Washington, D.C. (PB89-134159).
5. Ecology and Environment, Inc. Region X FIT Field Analytical Support Program Cost Analysis, Seattle, WA.
6. EPA and American Chemical Society. Fifth Annual Waste Testing and Quality Assurance Symposium, Washington, D.C., 1991.
7. NUS Corporation and Ecology and Environment, Inc. Field Investigation Team (FIT) Screening Methods and Mobile Laboratories Complimentary to Contract Laboratory Program (CLP), 1998.
8. EPA, SW-846, 1998, Washington, D.C.
9. U.S. EPA, *Test Methods for Evaluating Solid Wastes*, SW-846, 3rd ed., Office of Solid Waste and Emergency Response, Washington, D.C. November 1986.

chapter 7

Regulatory strategy

Regulatory issues are driven by: (1) defining the regulatory mechanism under which remediation is to occur (RCRA, CERCLA, LLRW, etc.), (2) defining the waste management requirements once remediation is enacted (disposal, transportation, etc.), and (3) developing the regulatory basis for establishing cleanup objectives.

This chapter provides an approach for assessing regulatory restrictions, requirements and options based on current or in-the-process regulatory initiatives. However, regulatory requirements and initiatives are evolving, changing entities. The discussion presented herein should be read with that in mind and should not be used as a shortcut for a more detailed, site-specific regulatory strategy that captures ongoing changes in regulatory requirements.

What is presented here is a philosophy for laying a foundation to attain maximum regulatory flexibility to achieve significant risk-based cleanup with minimal delay.

7.1   Macroengineering approach

In the U.S., a major challenge to the successful completion of hazardous and radioactive waste cleanup work is posed by overlapping, yet incomplete and sometimes inconsistent, regulatory requirements. In addition, state and federal statutes governing cleanup goals and procedures were intended to more easily apply to smaller and less complex facilities and oftentimes prove cumbersome for larger site remediations.

The problem of overlapping, inconsistent federal requirements was partially resolved in amendments to Resource Conservation and Recovery Act (RCRA) (1984) and Comprehensive Environmental Response, Compensation, and Liability Act (CERCLA) (1986). These amendments established "dominant" regulatory responsibility for cleanup at discrete facilities (e.g. steel mills, chemical plants, and landfills). However, the applicability

102   *Macroengineering: An environmental restoration management process*

(EPA 1986). The complete details of RCRA requirements for mixed wastes are still not fully determined. The full applicability of CERCLA requirements to a given site is not officially determined until the publication of a notice and supporting documentation that places the site on the National Priority List of Superfund sites.

Facilities that were closed before November 1980, or which are abandoned without recourse to a responsible owner or operator, usually are addressed under CERCLA. Brownfields or Voluntary Cleanup Program provisions of individual states have brought the two master remediation programs (RCRA and CERCLA) closer together.

Much of the new, small ($200,000 each) Brownfield grants have been given by EPA to facilitate interest and local political action to compel the states to provide nonfederal funds to start community-oriented actions to either locate willing developers or reluctant former owners to take positive action under favorable zoning and taxation conditions that can be granted at no direct cost by cities.

Active hazardous waste management facilities, or those closed since applicable RCRA regulations became effective in 1978, are usually subject to RCRA requirements, and this has become the *de facto* home of the Brownfields and voluntary cleanup initiatives. However, this neat separation of responsibilities and procedures does not work at all sites for several reasons.

The CERCLA and RCRA regulations include significantly different lists of regulated hazardous substances and waste constituents, respectively. The main examples are radionuclides, which are addressed under CERCLA, but not RCRA. A major example of incomplete requirements is the RCRA Corrective Action regulation, which will impose new national cleanup standards and procedures. These rules have been drafted by the EPA, but are not yet in effect. Also, some states apply additional control and cleanup requirements for historical disposal sites, including radionuclide contaminants and chemical agents.

Generally speaking, CERCLA, being operated under the broad provisions of the National Contingency Plan (NCP), is a less proscriptive program than RCRA. However, both RCRA and CERCLA have a wide array of technical and programmatic guidance and directive publications for reference. If anything, the plethora of guidance and directives requires close understanding of the regulatory process and how the best features of RCRA and CERCLA and their guidance and directives can be brought to bear in an organized and optimized manner to plan the cleanup.

The authors believe that a macroengineering approach can be constructed to be compliant with current applicable regulations. Furthermore, it is believed that new regulatory developments will increasingly encourage

macroengineering and were devised to be innovative and responsive to complex site conditions.

Key elements of a macroengineering regulatory strategy approach would be to:

- Establish a suitable spectrum of land- and groundwater-use alternatives.
- Select cleanup levels for water and soil that are consistent with the future use of specific portions of the site.
- If feasible, allow the use of natural processes for a site remediation while maintaining stringent institutional controls.
- Agree to the concept of a disposal facility (if necessary) that will be constructed on the site for final deposition of wastes removed during the cleanup of other areas.
- Establish a basis for consistent cleanup implementation based on the land-use alternative selected, regardless of the governing regulation. For example, an RCRA treatment, storage, and disposal (TSD) site in the TSD area would be subject to the same cleanup and procedures as a similar nearby CERCLA site.

Additional regulatory issues exist outside of the framework of the EPA and must also be addressed. These issues include:

- The applicability of Department of Transportation (DOT) and other off-site transportation requirements (labeling, packaging, etc.) to on-site waste movement
- The appropriateness and timing (if necessary) of a National Environmental Policy Act (NEPA) Environmental Impact Statement (EIS)
- The need to complete a siting study for a proposed TSD area disposal unit in accordance with other state and federal agency requirements

Oftentimes, major federal facility cleanups are governed by Tri-Party Agreements (EPA, state and federal agencies) and action plans that reflect determined efforts to pull together the sometimes inconsistent requirements of RCRA and CERCLA, as well as other state and federal statutes into a unified whole. In general, the most stringent requirements or particularly relevant guidance will be applied to cleanup actions.

Typically, a Tri-Party Agreement (TPA)-like document divides a site into separate operable units, with lead responsibility for regulatory review of

104   *Macroengineering: An environmental restoration management process*

procedures reflecting the fate and transport of each COC group of similar physical and chemical properties and characteristics.

However, typically the initial TPA efforts do not set future land-use plans, cleanup concentration limits or risk levels, or specify where and how wastes from the cleanup activities will be finally disposed. These have to be worked out as site/waste characterization data and public sentiments become more defined. Furthermore, these initial TPA documents in the past have typically been legal and bean-counting focused (i.e., identify all solid-waste management units [SWMUs]) with little emphasis on practical engineering implementation. Each operable unit RI/FS or RFI/CMS must provide adequate site characterization data and evaluate relevant remedial alternatives. Each remedial evaluation must identify, investigate, develop, and apply screening and acceptance criteria to a range of cleanup alternatives.

Figure 7.1 identifies the corrective measures evaluation criteria. Under the TPA setup, the potential for inconsistencies and unnecessary duplication of effort in dozens of operable unit investigations and remedial technology evaluations has tended to be high. Macroengineering addresses these potential inefficiencies in an increasingly comprehensive manner.

Under macroengineering, the action plan should include commitments to develop common procedures for environmental investigations and site characterization, data quality strategy, laboratory quality assurance, and other supporting documents. It should also include an impact study of major surface water bodies; methodology determining sitewide background values for soil and groundwater; and developing a standard risk assessment methodology. Preferably, the action plan should allow redefinition of operable units into larger aggregate areas for more detailed "scoping" prior to full remedial investigations and feasibility studies (RI/FS), set priorities for investigations on the most likely significant release sites (within aggregate areas), and further customize remedial decision processes. These suggestions are consistent with other comprehensive analyses of the Superfund RI/FS process (Johnson 1990; EPA 1990b) and the general site characterization and evaluation strategy commonly known as the *observational approach* (Myers 1990). The observational approach was developed through the works of Karl

Long-term effectiveness and performance	Reduction of toxicity, mobility, or volume of wastes	Short-term effectiveness	Implementability

Terzaghi (Harvard University soils engineer) and his younger colleague, Ralph B. Peck (University of Illinois soils, then geotechnical engineer). Their collaboration toward developing the observational approach began in the early 1940s and focused on tunneling operations in which quick response to unanticipated changes in condition was necessary. A similar circumstance exists in the environmental restoration setting.

The observational approach relies on the collection of limited site surface and subsurface data, here emphasizing both earth materials and contaminants. Cautious but comprehensive interpretation begins with a strong emphasis on applying previous experience and proven concepts. Uncertainties are sought and described and are carefully viewed in the total perspective with the expectation that site investigation and cleanup decisions will be refined and modified as investigation and actual cleanup work progresses. This approach avoids the common problem of studying a site until all uncertainties have been adequately reduced (a highly subjective determination) before proceeding with any cleanup work. This approach may not be appropriate in small, isolated sites with unique contaminants, but is very much applicable to large facilities that include many sites with similar geology, hydrology, disposal units, and contaminants. Under the observational approach, the public is served with evidence of ongoing progress, and the cleanup activities are highly tuned to ongoing collection of incoming data.

Geotechnical construction activities related to engineered construction, in general, have traditionally been conducted with on-site observation by geologists and geotechnical engineers, for the express purpose of verifying design assumptions and increasing the probability of matching of assumptions, design, and the ensuing construction. Such observation of construction-in-progress has been traditional in the U.S. for at least 80 yr.

In the environmental arena, employment of the observational approach has been in place since the late 80s in connection with expedited response actions (ERAs) at federal facility sites. These actions are intended to address perceived threats to public health or the environment in a relatively rapid fashion, as compared to the full RI/FS or RFI/CMS processes. The ERAs are similar to intermediate response measures (IRMs) of CERCLA and usually involve removal of low-volume, high-toxicity, high-concentration chemicals from unstable or free-to-migrate locations. They are targeted on very specific problems without becoming entangled in the overall uncertainties of final cleanup limits, detailed source area characterizations, and prediction of the extent of contaminant transport. The concept is wholly sound, especially when employed at source control to eliminate release of contaminants to earth media. These examples provide illustrations of application of the general process suggested for the measures studies (RFI/CMI), as the basis for determination of postclosure corrective actions and the basic requirements for gaining a clean-closure permit.

7.2   Special waste and remediation concerns

Radiation and unexploded ordnance (UXO) remediation have generated special concerns and regulatory approaches that need to be considered. They are briefly discussed herein.

7.2.1   Multi-agency radiation survey and site investigation manual

Radiation waste sites are additionally faced with identification and meeting considerations such as DOE/NRC decontamination and decommissioning activities (for inactive nuclear facilities). In the past, TPAs typically have not considered the DOE/NRC override requirements. Where decontamination and decommissioning work is required for relatively large surface structures, macroengineering has a significant forward-looking challenge. Recently, enhanced coordination on approach between EPA/DOE/NRC has been achieved through the development of the Multi-Agency Radiation Survey and Site Investigation Manual (MARSSIM).

MARSSIM is a consensus document designed and written by the federal agencies most involved with the radiation-site problem (DOE, EPA, NRC, etc.). MARSSIM is a technical document designed to guide and assist in the demonstration of compliance with cleanup standards as applied to radiation sites. For the most part, the basic questions of integrating some early remediation, to be followed by decontamination, the natural thought was to organize the action plan to identify IRMs, a process that naturally appeared due to evolving logic of the RCRA and CERCLA programs. MARSSIM takes the IRM approach a step further and provides a nationally consistent, scientifically rigorous approach to conducting performance-based surveys oriented towards dose- or risk-based regulations.

Figure 7.2 provides a comparison of the MARSSIM process compared to the CERCLA remedial process. As depicted in the figure, MARSSIM is a six-step process: (1) site identification, (2) historical site assessment, (3) scoping survey, (4) characterization survey, (5) real-time remedial action support survey, and (6) final-status survey.

The objectives of MARSSIM conform to the macroengineering approach, namely develop a conceptual sitewide model identifying potential contaminants and contaminated media, as well as establishing impacted and non-

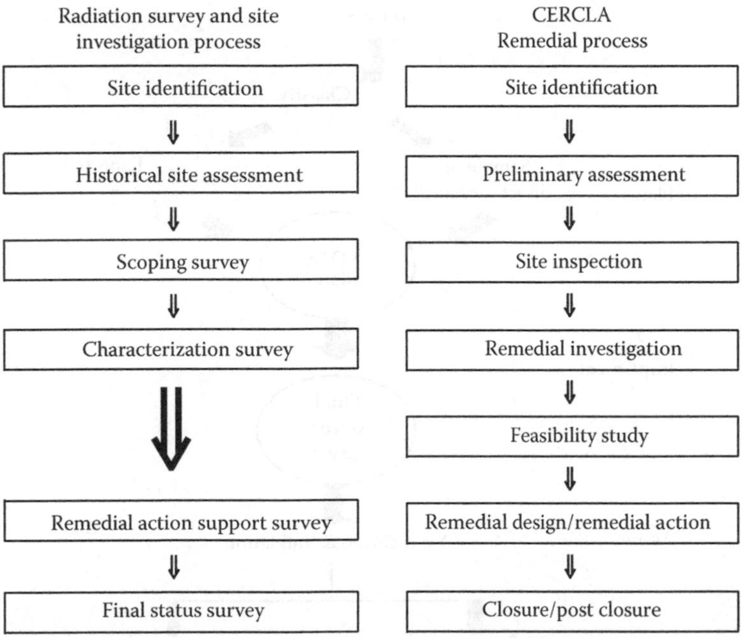

Radiation survey and site investigation process		CERCLA Remedial process
Site identification		Site identification
⇓		⇓
Historical site assessment		Preliminary assessment
⇓		⇓
Scoping survey		Site inspection
⇓		⇓
Characterization survey		Remedial investigation
⇓		⇓
		Feasibility study
		⇓
Remedial action support survey		Remedial design/remedial action
⇓		⇓
Final status survey		Closure/post closure

Figure 7.2 Comparison of MARSSIM and CERCLA remedial processes. Additional information on MARSSIM and regulatory compliance can be found in Appendix F of MARSSIM.

As depicted in Figure 7.3, MARSSIM provides comprehensive roadways on topics not covered by regulations, such as:

- How many physical samples to be taken and analyzed or direct measurements made to demonstrate compliance
- How to determine what physical sampling, analysis, or measurement method to use
- How to evaluate sample analysis and measurement data to determine if regulatory based criteria is met

As illustrated in Figure 7.4, MARSSIM is a data quality objective (DQO)-driven process that allows for decision making (even when there is uncertainty in the measurements) through statistical tests.

Conducting an appropriate statistical test requires some thoughtful decisions, such as:

108   *Macroengineering: An environmental restoration management process*

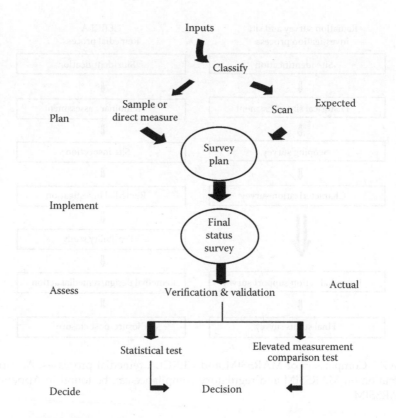

Figure 7.3 MARSSIM roadmap.

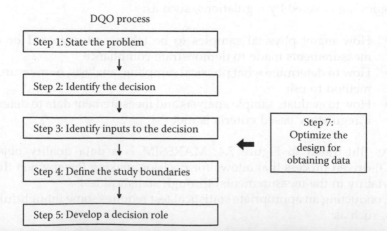

Chapter 7: Regulatory strategy 109

The primary goal is to establish the parameter of interest upon which cleanup decisions should be based. This is accomplished by establishing the derived concentration guideline level (DCGL). DCGL is a derived, radionuclide-specific activity concentration within a specified unit corresponding to a release criterion. The latter is a regulatory-limit cleanup standard expressed in terms of dose or risk. DCGLs are derived from activity-to-dose relationships through various exposure pathway scenarios.

The MARSSIM process can be used to establish the residual concentration below which it may be difficult or impossible to remediate. Furthermore, MARSSIM identifies the following two sample statistics as important to decision makers: (1) measures of central tendency (mean and median measurements) and (2) uncertainty (standard deviation of measurements).

Thus, decision parameter survey design involves trade-offs between increasing the number of measurements, using more precise measurement systems, creating more (and more homogeneous) survey units, and the ability to detect elevated areas.

The MARSSIM approach recognizes that each survey unit will have different physical characteristics that affect the scanning pattern and coverage selected for the survey. Typically, MARSSIM survey units are limited in size based on classification, exposure pathway, modeling assumptions, and site-specific conditions. Table 7.1 provides suggested survey unit areas based on area classifications.

The typical approach for the Class 1 areas is to conduct 100% surface scans over a number of data points established by statistical tests. Additional measurements, if necessary, should be conducted for small areas of elevated activity. The scan should be designed to deduct areas of elevated activity that would not be detected by a systematic pattern. As such, Class 1 area surveys should be based on a random-start statistical pattern.

Class 2 areas are typically subjected to 10 to 100% surface scans in either a systematic and/or judgmental pattern. The number of data points is established by statistical testing. Similar to Class 1, the Class 2 scan is designed to detect areas of elevated activity that would typically not be detected by a systematic pattern and is based on a random-start systematic pattern.

Table 7.1 Suggested Survey Unit Areas

Classification	Suggested area
Class 1	
Structures	Up to 100 m²
Land areas	Up to 2,000 m²
Class 2	

Class 3 area scans are based on a random/judgmental approach, scanning where experience tells the surveyor that contamination may exist, coupled with measurements performed at random with the number of data points being established by statistical testing.

In summary, the rationale for selecting a MARSSIM survey unit area should be developed using the DQO process and fully documented by data verification, validation, and quality assessments.

7.2.2   *Permitting and special demonstration requirements in radwaste*

Remedial actions performed entirely on site under CERCLA procedures are exempted from obtaining permits that might be required for similar actions in different circumstances. Section 121(e) of CERCLA provides this explicit exemption to avoid unnecessary administrative procedures and to expedite cleanup work. However, the substantive applicable or relevant and appropriate requirements (ARARs) of other local, state, and federal laws must be complied with. In broad terms, this means that statutory cleanup standards, criteria, or performance requirements must be met, even if some requirements are more stringent than those that are based on the CERCLA statute. However, this discussion will not attempt to include most of these additional laws. It is assumed that all or most of these requirements will be met by supplying adequate hazardous substance release control or treatment technology, protecting wildlife, archeological and cultural resources, etc.

The definition of *entirely on site* is important in this context. For many huge, macroengineering-type environmental cleanup, sites may have separate and distinct portions of the site placed on the CERCLA National Priority List. This might logically imply that transfer of wastes or contaminated soil from one listed area to another for disposal could be considered off-site disposal. This question is not completely answered in EPA guidance or other policy statements (EPA 1990a).

Guidance from EPA (EPA, 1988) indicates that *on site* will be considered to include the areal extent of contamination and all suitable areas in very close proximity to the contamination necessary for implementation of the response action. The CERCLA statute, regulations, and guidance are written to allow some flexibility in deciding the specific limits of a particular site. At a given site, the key determination appears to be whether an area proposed to be included in the remedial action "site" is necessary to the successful performance of the actions.

Chapter 7:   Regulatory strategy   111

permits under the state-administered National Pollutant Discharge Elimination System.

Off-site transport and disposal of untreated hazardous or mixed wastes are normally not allowed for CERCLA sites and are usually allowed only for trips to thermal decontamination facilities or specifically designed and permitted mixed-waste or low-level radwaste facilities. Even for RCRA compliance actions, transfers require manifesting and compliance with RCRA treatment standards or land disposal restrictions before disposal.

Disposal of mixed hazardous and low-level radioactive wastes in compliance with RCRA treatment standards or waivers is one of the main regulatory difficulties that cleanup plans must resolve. After May 8, 1992, high-level mixed wastes are specifically required to undergo vitrification treatment prior to disposal. In contrast, treatment processes that adequately manage both hazardous and radioactive components in most low-level mixed wastes are not specifically identified in regulations or anywhere else. The only low-level mixed wastes specifically identified in the RCRA land disposal restriction regulations in 40 CFR 268 are:

- Lead solids
- Elemental mercury
- Hydraulic oil (all contaminated with radioactive materials)

The treatment processes specified for these wastes are:

- Microencapsulation (e.g., surface coating)
- Amalgamation (with zinc or other appropriate materials)
- Incineration, respectively

EPA guidance for disposal of other low-level mixed wastes suggest that only the established treatment standard or process for the hazardous component will also apply to low-level mixed wastes, or that case-by-case variances or waivers must be obtained.

Oftentimes, proposed macroengineering approach plans for disposal of most of these low-level (including TRU) mixed wastes, or residues from basic separation and treatment of these wastes, will include straightforward landfilling. Fine soil particles washed from larger contaminated soil masses could be stored in large double-lined lagoons, whereas water from the treatment process is removed by evaporation. After completion of treatment and evaporation, the lagoons could also be closed in place as landfills.

These disposal plans, if done under CERCLA, do not necessarily include

of this requirement under a no-migration concept. Early resolution of this issue is of paramount importance.

It is worth noting that there is increased interest and support for dual-purpose land disposal facilities capable of handling both EPA regulated RCRA mixed-waste streams and NRC-controlled low-level radioactive waste streams. The design requirements of RCRA are bottom oriented (underlying liners, drainage sumps, et al.), whereas the NRC design approach is top focused (caps, etc.). A dual-purpose unit would undoubtedly have to incorporate both design philosophies.

For radwaste management units, a performance assessment must be conducted to evaluate and confirm the long-term viability of the design relative to regulatory criterion (and, more realistically, the actual longevity of the radionuclide contaminants left as a hazard. As part of the performance assessment, a strong emphasis should be given to assessing the containment designs vulnerability to degradation due to chemical attack (sulfates, chlorides, etc.) from constituents in the waste stream being managed.

Regulatory negotiations may be facilitated by considering bare-bones cleanup treatment and disposal scenarios, then working forward to define the minimum necessary refinements to develop a system which will provide adequate containment and long-term protection of public health and the environment. Resolution of these issues would be directly connected with the determination of required cleanup levels and future land and water-use plans.

7.2.3   UXO — range rule issues

Over the past several years, there has been increasing concerns raised over the UXO and hazardous chemical contamination at military ranges nationwide. Questions have been raised whether closed or transferred military ranges are being remediated in a manner consistent with accepted environmental or explosive safety standards and practices. To that end, EPA has been developing a Department of Defense (DOD) range rule. DOD has also supported EPA's efforts to define range rule remediation activities via its efforts to develop a range rule risk methodology and its implementation of a military munitions dialogue. Substantial progress has been made in improving the remediation process presented within the proposed rule, and toward developing a process to assess risks from UXO.

The common concerns that have typically been raised can be summarized as follows:

Chapter 7: Regulatory strategy 113

- Poor coordination and information distribution, as well as incomplete UXO and contaminant information on both a site-specific and national basis
- Remedy selection and implementation problems when large-scale UXO cleanup actions are run by default as CERCLA-like removal actions or RCRA emergency situations
- General concerns over property transfers in which UXO may be remaining.

It is worth noting that most environmental restoration activities at DOD sites up to 1997 were not directed at UXO assessment and response but were directed towards open-burning and disposal grounds and nonexplosive chemical contamination. UXO in potential firing areas were not included in the realm of the potential cleanup or even identified as areas of concern. However, in 1998, the Army tentatively agreed to evaluate areas known or suspected to be contaminated with UXO, using the Sitestats/Gridstats analytical methods. Thus, there has been an increasing tendency to use statistical grid sampling methods for UXO investigations. Although statistical grid sampling may yield information, extrapolation of these results may lead to inappropriate decisions, because it assumes a relatively uniform distribution of UXO that is typically not the case at military ranges.

The proposed range rule process is heavily dependent upon accurate informed risk management decision making that emphasizes reducing short-term risks and setting the stage to achieve long-term risk-reduction goals that are not met by statistical grid sampling and surface clearance methods now in place. In fact, DDESB standards recommend much more conservative clearance for the residential land-use option that is typically the public remedial goal.

Generally, sites have not been applying the best available technologies to assess and remediate UXO. In most cases, there appears to be a default to the traditional methods known as *Mag and Flag* vs. Brookes geophysical survey methods. Also, in those cases in which UXO investigations have been performed, the general approach has been to limit the investigations to known ranges/UXO sites only, despite site information that suggests more extensive UXO problems.

Last, it is felt that there is an over-reliance on institutional controls as the principle remedy component or as the only remedy to ensure protectiveness in the last resort. The institutional controls may not be adequately defined with roles and responsibilities left unclear and ultimately may fail to prevent future incidents of UXO being encountered.

114   *Macroengineering: An environmental restoration management process*

around 1983, was eventually borrowed and applied to the RCRA corrective action program, as the SWMU concept.

Each unit was considered a unique environmental restoration problem that needs a thorough investigation, followed by a remedy selection that involves a process of formally considering a range of potential solutions. Usually, the nature and extent of site contamination was characterized by an RI, followed by a detailed evaluation and comparative analysis of several remediation alternatives in a feasibility study, with one of the alternatives being proposed to the public in a proposed plan. The latter is then designated the alternative in a record of decision (ROD), if its suitability is adequately supported by technical data and arguments. With the publication of the Superfund ROD, the remedial design develops exact technical specification and relevant construction dimension details for competitive bidding among cleanup contractors. Subsequent cleanup takes place during what is referred to as a *remedial action phase*. The whole CERCLA remedy selection process has historically taken 5 to 7 yr for each site (unit).

Thus, in the traditional remedy selection process, several alternative remedies are evaluated for each site (unit) and site characterization is essentially complete before final remedy selection. As discussed in Chapter 1, if the remedy selection is based on inadequate data or involves a technology with limited range, unexpected site characteristics might render the selected remedy ineffective, yet regulatory speaking, the site management is in inflexible position to quickly adapt to the revised circumstances.

The presumptive remedy concept was introduced by EPA's Office of Solid Waste & Emergency Response in September 1993 as a means of applying common-sense remedy selection to the simplest cases of uncontrolled hazardous waste and, given the nature of Brownfield remediation efforts, has broad application to these remediation efforts. The approach calls for EPA to presume that a remedial technology is appropriate where voluminous treatability data have already confirmed the overall effectiveness of a particular remedial technology, as applied to specific hazardous wastes. In these situations, EPA does not require development and comparison of multiple alternatives to be competitively evaluated for the selection of the best remedy, but accepts the applicability of the remedial technology based on previous application results in other similar situations.

The plug-in or tiered risk approach is a subset of the presumptive remedy concept, developed largely by state environmental agencies as a means of responding to small-site spills and discharges of specific contaminants, generally petroleum hydrocarbons related to the leaking underground storage tank program. Although this concept's small-site emphasis does not inherently fit within the large-scale macroengineering orientation at a glance, the

Chapter 7: Regulatory strategy   *115*

committee of its members. These selection mechanisms have grown onto three tiers, beginning with the look-up tables (Tier One), followed by use of known or reasonably-proven relationships of contaminant concentrations as they relate to human health (Tier Two), and the traditional development of a risk-based scenario of identified receptors being subjected to tailored exposures to candidate COCs for the particular site (Tier Three). Tier Three represents the standard, site- and waste-specific risk assessment methodology developed for CERCLA sites in about 1984.

The plug-in approach also allows multiple separate subsites within a larger site to utilize the same remedy at different times via an expedited, paperwork-streamlined fashion. Under the plug-in approach, EPA accepts a standard remedy that is applicable to a given set of conditions, rather than requiring separate regulatory support packages and documentation for each specific subsite. For this streamlining option, EPA establishes a set of criteria for determining where those conditions are applicable to the overall site. This allows for effective use of decision tree framework management in the field. Furthermore, under the "plug-in approach," subsites can by fully characterized by individual cleanup types [SWMUs and corrective action units (CAUs)] specified in the ROD, and utilize real-time characterization and observational approach consistent with the macroengineering environmental restoration management method. Thus, based on the preestablished process and decision criteria established by the ROD, EPA can facilitate making subsite-specific determinations that "plug-in" subsites to the remedy. The plug-in approach affords the site and the regulator with flexibility to address unforeseen circumstances without going through a repetition and time-consuming process for reselecting the same remedy at each subsite.

Let us work through an example. A given site has zones of VOCs in soils, separated by large zones of uncontaminated surficial soil. In most cases, VOC-contaminants correspond to certain units' locations, but hydraulically downgradient contaminant transport has been such that the VOC-contaminated soils span below a cluster of units. EPA has the option to consider the one continuous zone of VOC soil contamination and the associated units a subsite that, in turn, makes it a candidate for a plug-in approach for source removal. Considering the subsurface contamination within the vadose (saturated zone below the groundwater surface), the contamination plume becomes one VOC contamination problem. A single remedial action technology, for instance, one vacuum extraction cleanup system could then be applied. In this example, the plug-in remedy identifies vacuum extraction as the standard remedial action technology. Thus, the ROD does not select a remedial action for a specific subsite, but instead designates a remedial action technology as applicable to

The **Existing Site Profile** specifies the range of common conditions among the potential subsites defined in terms of the various physical and contaminant parameters that could have an impact on the remedial alternative effectiveness.

The **Presumed Remedial Alternative** is the technology action that will be taken at the subsites that meet the Remedy Profile and Plug-In Criteria.

The **Remedy Profile** is the range of conditions that the Presumed Remedial Alternative can handle. Note, the Remedy Profile can be expanded via technical enhancement in the Presumed Remedial Alternative pre-specified in the ROD.

Technical Enhancement to the Presumed Remedial Alternative may be appropriate in three situations: (1) to widen the Remedy Profile; (2) to make the Presumed Remedial Alternative more efficient; or (3) to meet an ARAR condition struck with the other subsite criteria.

The Plug-In Criteria must be based on potential health threats that serve as the standard for EPA to determine whether an action is necessary.

The Plug-In Decision Point specifies the conditions under which a Presumed Remedial Alternative is applicable. There are two conditions that a subsite must meet to be plugged in: (1) the subsite must exhibit conditions that fall within the Remedy Profile; and (2) the contamination at the subsite must exceed the Plug-In Criteria. Note, from a site management standpoint it is critical that the Plug-In Decision Point clearly reflects the boundary of site characteristic variance for which a given technology is feasible.

Figure 7.5 Plug-in approach — ROD elements.

reduction in the degree of characterization needed to justify later, similar actions at other on-site SWMUs or CAUs. Also, schedule and cost advantages may be achieved by performing the later RI and remedial action concurrently, as well as the advantage of moving forward on multiple sites in a concurrent economy-of-scale fashion.

Relative to the plug-in criteria in our example, VOCs may leak downward and enter groundwater, or they may volatize upward and be inhaled near the ground surface. The plug-in criteria, in effect, should set separate limits on the levels of VOCs that may reach the groundwater and levels of

Thus, the plug-in approach incorporates all the basic components of the traditional Superfund process, but streamlines the process to minimize redundancy and optimize the sequence of activities.

Figure 7.5 presents the typical elements relative to a plug-in approach ROD and how these elements ensure appropriate application of the designated remedy to subsites.

It should also be noted that the voluntary cleanup program of the individual states, coming into prominence since 1989, has also instilled a clear-the-decks attitude, which favors consideration of innovative interpretations and approaches to cleaning up previously contaminated industrial sites.

7.4   Applicable or relevant and appropriate requirements

ARARs begin with those requirements technically defined according to the Comprehensive Environmental Response, Compensation and Liability Act (CERCLA). ARARs may also be more broadly understood as factors important to the accomplishment of a task, or achievement of a goal.

The technical ARAR definition is specified in Section 121(d)(2)(A) of CERCLA. ARARs include promulgated federal standards, requirements, criteria, or limitations that are legally applicable or relevant and appropriate to the site or situation in question. State requirements may be ARARs if they have legal standing and are at least as stringent as federal requirements. The primary EPA guidance for ARARs is the two-volume *CERCLA Compliance with Other Laws Manual.* Approximately 35 additional CERCLA directives, fact sheets, and guidances (see Figure 7.6) may be used in identifying contaminant-, location-, or action-specific ARARs.

ARAR precedents established in previous RODs for similar sites or situations may represent important regulatory strategy development. ROD information should be retrieved and examined to determine its precedence value and, where appropriate, apply relevant precedents (as warranted) to maintain consistency. The ROD system (RODS) database at EPA HQ can be queried to identify final RODs by keywords, and then additional detailed information (e.g., ARAR documentation) from specific RODs can be ordered or examined in a regional library or Superfund record center.

The technical approach to ARAR documentation is to not only determine the details and legal status of potential requirements, but to identify those that may have the greatest impact on cleanup or remediation plans if they are not met. In addition, uncertainties regarding applicability or relevance of some potential ARARs (a common circumstance) are explicitly identified.

ARARs are specifically examined to form boundary of uncertainties and

---

CERCLA Compliance With Other Laws Manual Part I (Interim Final)

CERCLA Compliance With Other Laws Manual Part II: Clean Air Act and Other Environmental Statutes and State Requirements

Discharge of Waste Water from CERCLA Sites into Publicly Owned Treatment Works

Consideration of ARARs During Removal Actions

Superfund LDR Guide #1: Overview of RCRA Land Disposal Restrictions (LDRs)

Analysis of Treatability Data for Soil and Debris: Evaluation of Land Ban Impact on Use of Superfund Treatment Technologies

ARARs Qs and As: Compliance with Federal Water Quality Criteria

ARARs Qs and As: Compliance with New SDWA National Primary Drinking Water Regulations for Organic and Inorganic Chemicals

ARARs Qs and As: Compliance with the Toxicity Characteristics Rule: Part 1

ARARs Qs and As: The Fund-Balancing Waiver

ARARs Qs and As: General Policy, RCRA, CWA, SDWA, Post-ROD Information and Contingency Waivers

ARARs Qs and As: State Groundwater Anti-Degradation Issues

Applicability of Land Disposal Restrictions to RCRA & CERCLA Groundwater Treatment Reinjection, Superfund Management

CERCLA Compliance with Other Laws Manual: CERCLA Compliance with the Clean Water Act (CWA) and the Safe Water Drinking Act

CERCLA Compliance with Other Laws Manual: CERCLA Compliance with the State Requirements

CERCLA Compliance with Other Laws Manual: Guide to Manual

CERCLA Compliance with Other Laws Manual: Overview of Applicable or Relevant and Appropriate Requirements (ARARs) - Focus on ARAR Waivers

CERCLA Compliance with Other Laws Manual, Part 1 (Interim Final)

CERCLA Compliance with Other Laws Manual, Part 2: Clean Air and Other Environmental Statutes and State Requirements

CERCLA Compliance With Other Laws Manual: RCRA (Resource Conversation and Recovery Act) ARARs (Applicable or Relevant and Appropriate Requirements) - Focus on Closure Requirements

CERCLA Compliance With Other Laws Manual, Summary of Part 2: CAA, TSCA, and Other Statutes

CERCLA Off-Site Policy: Eligibility of Facilities in Assessment Monitoring

CERCLA Off-Site Policy: Providing Notice of Facilities

(a)

Figure 7.6 CERCLA, ARAR directives, fact sheets, and guidances.

7.5 *Review of RCRA corrective action regulatory initiatives*

Chapter 7: Regulatory strategy   119

Consideration of RCRA Requirements in Performing CERCLA Responses at Mining Waste Sites, Delegations of Authority Under the Federal Water Pollution Control Act (FWPCA) which are Applicable to the Superfund Program

Discharge of Waste Water from CERCLA Sites into Publicly-Owned Treatment Works (POTWs)

Interim RCRA/CERCLA Guidance on Non-Contiguous Sites and On-Site Management of Waste and Treatment Residues

Land Disposal Restrictions as Relevant and Appropriate Requirements for CERCLA-Contaminated Soil and Debris

Notification of Out-of-State Shipments of Superfund Site Wastes

Policy for Superfund Compliance with the RCRA Land Disposal Restrictions

Summary of Notification of Out-of-State Shipments of Superfund Site Waste

Superfund LDR Guide #1: Overview of RCRA Land Disposal Restrictions

Superfund LDR Guide #2: Complying with the California List Restrictions Under Land Disposal Restrictions (LDRs)

Superfund LDR Guide #3: Treatment Standards and Minimum Technology Requirements Under Land Disposal Restrictions (LDRs)

Superfund LDR Guide #4: Complying with the Hammer Restrictions Under Land Disposal Restrictions (LDRs)

Superfund LDR Guide #5: Determining When Land Disposal Restrictions (LDRs) are Applicable to CERCLA Response Actions

Complying with the California List Restrictions Under Land Disposal Restrictions (LDRs)

Superfund LDR Guide #6A (Second Edition): Obtaining a Soil and Debris Treatability Variance for Remedial Actions

Superfund LDR Guide #6B: Obtaining a Soil and Debris Treatability Variance for Removal Actions

Superfund LDR Guide #7: Determining When Land Disposal Restrictions (LDR) are <u>Relevant</u> and <u>Appropriate</u> to CERCLA Response Actions

Superfund LDR Guide #8: Compliance with Third Requirements Under the LDRs

A Guide to the Delisting of RCRA Wastes for Superfund Remedial Responses

Superfund LDR Guide #10: Guide to Obtaining No Migration Variances for CERCLA Remedial Actions

CERCLA Compliance with the RCRA Toxicity with the RCRA Toxicity Characteristics (TC) Rule: Part 2

Superfund LDR Guide to RCRA Management Requirements for Mineral Processing Wastes

(b)

Figure 7.6 Continued.

the proposed Subpart S regulations, particularly those that pertained to stabilization favor a more aggressive environmental restoration management

120   Macroengineering: An environmental restoration management process

of a listed waste and a solid waste must be managed as a hazardous waste
unless it has been delisted. The current regulations also include a
"derived-from" rule (40 CFR 261.3(c)(2)(i)(d)(2)). The "derived-from" rule
provides that any solid waste that is generated in treatment, storage, or
disposal of listed wastes must be managed as a hazardous waste unless it
also is delisted. In addition, the "contained-in" rule (40 CFR 261.3(c)(2)(i))
states that any waste that contains a listed waste such as rags, solvent
materials, and clothing must also be managed as a hazardous waste.

The "mixture," "derived-from," and "contained-in" rules define a waste
as "hazardous" under RCRA without regard to the concentration of hazardous constituents in the waste itself or the mobility of those constituents. As
a result, large volumes of waste have historically been forced into regulated
status despite posing little or no risk to human health or environment.

The purpose of the HWIR is to set forth alternatives that will remove
low-risk wastes from unnecessarily stringent management requirements.

The two alternatives being examined under the HWIR are:

- *Concentration-based exemption criteria* (CBEC) that would define when
  a particular waste exits the RCRA Subtitle C waste management
  scheme. The exemption criteria would be based on a single concentration limit for each toxicant in the listed waste and could be based
  on risk factors such as health-based numbers (HBN), best available
  technology (BAT), or a combination of the two. This approach would
  remove from Subtitle C hazardous waste management any waste that
  is below those concentration levels.
- *Enhanced characteristic option* (ECHO) that proposes expanding the
  present set of four characteristics or possibly expanding the scope of
  the toxicity characteristic, and establishing characteristic concentration levels for all Appendix VIII hazardous constituents.

There are other aspects of the EPA's HWIR proposal that may be of
interest relative to the implementability of a macroengineering environmental restoration approach. The HWIR proposal introduced a concept called
"contingent management" that would exempt waste within certain hazardous constituent concentration ranges from Subtitle C regulation if the wastes
were managed in a prescribed manner such as placement in a lined landfill.
Contingent management could allow wastes with hazardous constituent
concentration levels that exceed the exemption numbers, but which are
shown not to pose a serious risk to human health or the environment. Such
wastes would than be exempt from Subtitle C regulation, upon approval of
proposed management techniques.

health-based number. The second tier would allow wastes to exit Subtitle C if their TCLP extract is between 10 and 100 times the health-based number; but, in this case, the waste must be managed according to Subtitle D landfill regulations (40 CFR Part 258) that require a single composite liner or equivalent. Wastes with the TCLP extract 100 times the health-based number would be subject to full Subtitle C landfill standards.

The proposed contingent management option under ECHO would be based on site-specific factors, such as liner system, size of landfill, precipitation, soil type, and proximity of drinking wells. These proposed options would take into account toxicity-characteristic limits based on health-based limits, as well as dilution and attenuation factors. The latter are plugged into the EPA composite model for landfills (EPACML). In this manner, higher thresholds above the exemption concentration levels could be established.

EPA is also examining the interaction between the proposed CEBC and ECHO approaches and the RCRA land disposal restrictions (LDR). The exemption criteria being proposed could become the minimized threat standard required by the LDR. Thus, wastes treated to satisfy the exemption criteria could be land disposed without further treatment under this interpretation.

HWIR may also impact the contaminated media regulated under the contained-in rule. Under one proposed approach, contaminated media would be addressed in a manner similar to any other Subtitle C waste and thus provide considerable relief to those undertaking RCRA corrective actions and/or CERCLA remediation that deals with minimally contaminated media.

The second key proposed federal regulatory position is the proposed Subpart D proposals and the potential implication of the "stabilization" concept. Stabilization is a new strategy that EPA has been implementing since the beginning of 1992. The purpose of stabilization is to achieve more efficient and quicker environmental results at RCRA treatment, storage, and disposal facilities requiring corrective action. Stabilization emphasizes controlling releases and preventing the further spread of contaminants. For the most part, it is expected that stabilization will involve excavation and on-site management of contaminated soil, sludge, and other wastes that are subject to RCRA Subpart C hazardous waste regulations.

As part of the process of proposing the new Subpart S corrective action regulations, EPA has promulgated the CAMU to facilitate effective remedial actions. The CAMU concept is similar to the CERCLA concept of "area of contamination." The latter allows for broad areas of contamination that often include specific subunits to be considered as a single land disposal unit for

EPA is recognizing that the strict application of RCRA LDRs and minimum technology requirements (MTRs) may limit or constrain remedial options that would be available, as well as affect the volumes of materials that are to be managed. Unfortunately, strict application of LDRs and possible MTR requirements could delay remediation, while providing little (if any) additional environmental protection for the site.

Subpart S and CAMU rules have not yet been finalized. Existing regulatory authority may, however, allow some implementation of this type of approach in site remediation and stabilization actions, in which it can promote effective and expeditious remedial solutions. The main thrust of this line of reasoning is to designate an area of contamination as a "mega-landfill." This will require that the unit comply with certain RCRA requirements that are applicable to landfills. Those requirements, however, will differ depending on whether the mega-landfill is considered to be an existing nonregulated landfill or a regulated hazardous waste landfill. This is determined by the regulatory status of the units or areas that are included as parts of the mega-landfill.

In one case, the mega-landfill would be considered an existing nonregulated landfill if all of the SWMUs within its boundaries are not regulated as newly permitted hazardous waste units (HWMUs) under RCRA. In this case, the mega-landfill would not be considered to contain newly generated hazardous wastes and would therefore not be subject to RCRA Part 264 or 265, design and operating requirements for hazardous waste landfills.

Furthermore, by designating the whole area as a single landfill, EPA can approve movement and consolidation of the hazardous wastes and soils contaminated with hazardous wastes and contaminated soils within the unit boundary without triggering the LDRs and MTRs. The mega-landfill could be subject to a number of regulatory restrictions. It cannot receive hazardous wastes from other units outside its CAMU boundary (either on-site or off-site), nor would hazardous waste treatment be allowed to occur at the mega-landfill site, nor can hazardous waste be removed from the mega-landfill and returned back after treatment. In all three instances, the mega-landfill would have to come under 264 and 265 regulations.

In the second instance, where the mega-landfill is considered a regulated landfill when accepting wastes generated or disposed within its own CAMU boundary, however, MTRs would not necessarily apply to the newly designated regulated mega-landfill.

The question of RCRA stabilization strategy is still open as to how such can be integrated with a Superfund effort. The environmental priority initiative (EPI) is an integrated RCRA/Superfund effort used to identify and evaluate contaminated sites that present the greatest risk to human health

It would appear that macroengineering-sized sites would obviously be a likely site for stabilization efforts. Stabilization treatment, using the CAMU concept, would obviously be compatible with the Superfund "area of contamination" concept that in fact it is modeled after.

7.5.1   Corrective action management units

EPA has recently issued its final decision concerning the CAMU and temporary unit (TU) portions of the proposed Subpart S corrective action regulations [first published in the *Federal Register* on July 27, 1990 (55FR 30798-30884)]. The Subpart S rule contains several key remediation waste management provisions that were designed to reduce or eliminate certain waste management requirements of the current RCRA Subtitle C regulations. Through these changes, EPA has expedited promulgation of these key provisions. The proposed ruling incorporates cleanup guidelines and strategies of which the CAMU and TU are integral concepts. CAMU and TU corrective action provisions under Subtitle C (CAMU/TU Rule) were signed by the administrator on January 14, 1993, published in the *Federal Register* on February 16, 1993 (58 FR 8658), and are effective from April 1993.

The CAMU/TU rule defines three key terms integral to the CAMU concept. The rule defines a *Corrective Action Management Unit* as "a contiguous area within a facility as designated by the regional administrator for the purpose of implementing corrective action requirements of this subpart, which is contaminated by hazardous wastes (including hazardous constituents), and which may contain discrete, engineered land-based subunits." The definition specifies that CAMUs may be used for corrective actions under Section 3008(h) orders, as well as at permitted facilities under Section 3004(u). Also, the definition specifies that CAMUs are to be used only for the purposes of managing remediation wastes.

The definition of CAMU specifies that only remediation wastes will be managed in a CAMU. The rule then defines what will be considered remediation wastes in designating and implementing CAMU at a facility. *Remediation wastes* are defined as "...all solid and hazardous wastes, and all media (including groundwater, surface water, soils, and sediments) and debris that contain listed hazardous wastes, or which themselves exhibit a hazardous waste characteristic, that are managed at a facility for the purpose of implementing corrective action requirements under §264.101 and RCRA 3008(h)." For a given facility, remediation wastes may originate only from within the facility boundary, but may include waste managed in implementing RCRA Section 3004(v) or Section 3008(h) for release beyond the facility boundary." The definition includes wastes generated as part of site investigations (i.e.

124   *Macroengineering: An environmental restoration management process*

HSWA-mandated corrective actions. Both CAMUs and TUs are restricted to managing wastes that are generated in implementing corrective action at a facility with either existing RCRA permits, or to those operating under interim status.

With the definitions of CAMUs, remediation wastes, and facility clearly stated, the rule clarifies several administrative/authoritative concepts. Section 264.552(a)(1) and (2) specify the essential regulatory basis for the CAMU as follows:

- Placement of remediation wastes into a CAMU does not constitute land disposal of hazardous wastes.
- Consolidation or placement of remediation wastes into or within a CAMU does not constitute creation of a unit subject to MTRs of Subtitle C.

This section removes the regulatory disincentives to more protective corrective actions, because land-ban restrictions (LDRs), best-demonstrated available technologies (BDAT), and MTRs are no longer applicable for placement of remediation wastes into or within a CAMU. EPA believes that by providing an alternative to the Subtitle C requirements, CAMUs provide greater waste management flexibility and, therefore, enhance the ability to select and implement effective, protective, reliable, and cost-effective remedies for RCRA facilities. The basis for the decision is that the original intent of the Subtitle C (HSWA) provisions was to prevent new releases from the management of "as-generated" hazardous wastes. Subtitle C requirements, when applied to "as-generated" wastes, are meant to ensure that wastes are handled in accordance with the stringent national standards, and ultimately lead to the generation of less hazardous waste because of the cost impacts of these stringent standards. If applied to corrective action, Subtitle C creates a strong disincentive for treatment or consolidation of remediation wastes. EPA predicts that the CAMU/TU rule will result in more on-site waste management, lesser reliance on incineration, greater reliance on innovative technologies, and lower incidence of capping the waste in place without treatment.

The rule emphasizes that a CAMU is considered a land-based unit and as such cannot include incinerators, tanks, treatment units, etc. With the exception of incinerators (and other thermal treatment processes), tanks and other treatment units can be considered to be TUs and can be located within the physical boundaries of the CAMU. They will have to comply with the requirements for TUs set forth in the rule. The regional administrator may designate different land-based waste management techniques within a CAMU and has the authority to specify specific standards for each area under Part 264 or 265.

are landfills, surface impoundments, waste piles, and land treatment units that received hazardous wastes after July 26, 1982. These units are subject to full Subtitle C design, operating, closure, postclosure, and financial responsibility requirements under Subparts F, G, H, and the unit-specific requirements of Part 264 and Part 265. Although a regulated unit may incorporate into a CAMU, it is subject to the following limitations:

- "Only closed or closing units (i.e., those units required to begin the closure process under §264.113 or §265.113) would be able to be so designated. Operating regulated units, including regulated units continuing to operate under delay or closure provisions (in §264.113 or §265.113) would not be eligible for designation as CAMUs."
- "The regional administrator will have the authority to designate a regulated unit as a CAMU, or as a part of a larger CAMU, only if doing so will enhance implementation of an effective, protective, and reliable remedy for the facility."

When a regulated unit is designated part of a CAMU, the rule requires that applicable portions of Part 264 or Part 265 groundwater monitoring, closure, postclosure, and financial responsibility requirements continue to apply to the unit as before. However, inclusion of a regulated unit into a larger CAMU would not cause the entire CAMU to become subject to the standards applicable to the regulated unit. This does not necessarily release the owner/operator from groundwater monitoring, closure, postclosure, and financial responsibility requirements for the CAMU. It just makes the distinction between these requirements for the regulated unit and the CAMU. Specific changes in the compliance agreement need to be made to incorporate overall aspects of the CAMU postclosure performance. The negotiated provisions include groundwater monitoring, closure, postclosure, and financial responsibility requirements.

Once the decision has been made to designate a CAMU, it must be evaluated against seven criteria. Section 264.552(f) requires the regional administrator to document the rationale for designating a CAMU and explain the basis for designation in permit or order. This rationale or basis must include an evaluation against the seven criteria. The seven criteria are as follows:

- The CAMU must facilitate the implementation of a reliable, effective, protective, and cost-effective remedy. EPA clearly states that the emphasis of these criteria must be on protectiveness, and that if the use of a CAMU "will not result in remediation activities with these qualities," it will not be designated by the regional administrator.
- Remediation waste management associated with CAMUs cannot cro-

126   *Macroengineering: An environmental restoration management process*

EPA clearly states that a qualitative approach to risk assessment "will generally be sufficient unless the regional administrator deems more quantitative data are necessary."

• The regional administrator must ensure that any land area of a facility that is not already contaminated (i.e., where there is no soil contamination or where wastes are not already located) will be included within a CAMU only if remediation waste management at such an area will, in the regional administrator's opinion, be more protective than management of such wastes at contaminated areas of the facility. Again, EPA states that their intent is not to use "formal risk assessments or other quantitative analyses" to support protectiveness decisions. Furthermore, the EPA states that inclusion of clean areas within CAMUs "will be allowed only if doing so is consistent with the overall remedial objective of the CAMU and will, in fact, be more protective than management of such wastes at contaminated areas of the facility."

• Areas within a CAMU where wastes will remain in place after closure of the CAMU are to be managed and contained so as to minimize future releases, to the extent practicable. The EPA states that any CAMU decision must consider, as a primary objective, the long-term (i.e., postclosure) reliability and effectiveness of CAMU-related remedial actions.

• The CAMU will expedite the timing of remedy implementation, when appropriate and practicable. The EPA clearly states that "the regional administrator is encouraged to utilize CAMUs if they will assist in eliminating unnecessary delays and will encourage faster pace to remediation."

• The CAMU-related remedial action should incorporate, as appropriate, treatment technologies (including innovative technologies) to enhance the long-term effectiveness of remedial actions at the facility by reducing the toxicity, mobility, or volume of wastes that will remain in place after closure of the CAMU. This criterion is analogous to the preference under CERCLA for treatment-based remedies and reflects the agency's general preference for permanent reduction in the overall degree of risk posed by wastes and for source-control remedies that involve treatment. However, the agency clearly states that "this criterion does not preclude remedial actions that do not employ treatment, as long as they are capable of ensuring long-term effectiveness."

• The CAMU will minimize the land area of the facility upon which wastes will remain in place after closure, to the extent practicable. The EPA's intent is "to promote consolidation of remediation wastes into smaller, discrete areas of the facility, that are suitable as long-term repositories for the wastes, and which can be effectively managed

Chapter 7: Regulatory strategy 127

Section §264.552(d) requires the owner/operator to supply information sufficient for the regional administrator to assess the decision criteria specified earlier. The rule does not explicitly require submission of an RFA, RFI, or CMS as part of the CAMU designation. Rather, the EPA intended to simply express the general authority under 3004(u) and 3008(h) to require information from the owner/operator necessary to support corrective action implementation decisions. The format and method of submission of this information is meant to be flexible and, as such, fits well to the macroengineering concept. Supporting information could be submitted at the same time or before any application for permit modification related to CAMU designation.

Once the decision is made to utilize a CAMU in a corrective action and after the previously described supporting information has been developed, the CAMU may be approved under an agency-initiated permit modification (§270.41) or according to the permit modification procedures of §270.42 for owner/operator-initiated modification. In the case of owner/operator-initiated modifications, the supporting information, including the assessment/evaluation of the CAMU against the seven criteria, could be submitted as an attachment to the application. Also when incorporation of a CAMU is initiated by an owner/operator, it will generally be approved (or disapproved) according to the Class III permit modification procedures. The following requirements for the CAMU will be specified in the permit:

- The areal extent and configuration of the CAMU.
- How remediation wastes will be managed in or as part of the CAMU, including specification of design and operating and closure requirements.
- Groundwater-monitoring requirements for each CAMU. (The rule does not provide specific, detailed groundwater-monitoring requirements addressing the numerous technical elements of installing and operating an effective groundwater-monitoring system. Rather, the rule establishes a general standard of performance for the systems; detailed specifications or performance standards will be specified in the permit, based on site-specific information and conditions.)
- Closure and postclosure requirements for each CAMU. (Similar to groundwater-monitoring requirements, the rule does not provide specific detailed closure and postclosure requirements. Rather, it seeks to set general performance standards, with detailed specifications and performance standards set based upon site conditions and the waste management activities being conducted at the CAMU.)

128   *Macroengineering: An environmental restoration management process*

7.5.2 *Temporary units*

EPA also recognized that the technical requirements specified in 40 CFR Part 264 regulations may be inappropriate for short-term management of wastes during corrective action. In addition, EPA clearly states in the preamble to the proposed Subpart S rule that in many cases applying these stringent Part 264 standards, which are designed to ensure adequate protection for long-term management of hazardous wastes, would be unnecessary from a technical standpoint, as well as counterproductive in many cases for short-term management of wastes. The CAMU/TU rule (55FR 30798-30884) finalizes the TU provisions of the proposed Subpart S rule with minor revisions. The CAMU/TU rule allows tanks and container storage units used for treatment or storage of remediation wastes to be eligible for designation as TUs. EPA states in the preamble to the CAMU/TU rule that "the site-specific review and oversight that is provided in the context of investigating and making remedial decisions for corrective action allows the agency to ensure protection of human health and the environment for short-term operation of units that may not meet the full set of standards specified for long-term use of such units under current RCRA regulations."

Early site environmental restoration planning must be proactive and must take note of several limitations to the type units that can be designated as TUs. Land-based units such as waste piles will only be addressed under the CAMU provisions of the rule. In the preamble to the rule, EPA says that "land-based waste management activities are more effectively addressed" under the CAMU provisions. The TU rule also limits the level of technology allowable for a TU. The preamble to the Subpart S rule states that "this provision for temporary units could apply to any unit used during corrective action, except incinerators and no-tank thermal treatment units (e.g., pyrolysis units)." There is a preference for low-level technology for TUs. There is a feeling that the complexity of high-tech treatment devices creates a higher level of public concern about their operation. Furthermore, the rule only allows the regional administrator to relax or modify the technical standards for TUs, not performance standards. Releases to the environment such as those contained in 40 CFR, Part 269 concerning air emissions, cannot be modified. Lastly, the TU rule specifically restricts the location of a TU to the facility where the corrective action is taking place.

Similar to the CAMU provisions, the TU provisions of the rule specify seven decision factors/criteria for the regional administrator to consider when designating a TU. These criteria are as follows:

6. Hydrogeological and other relevant environmental conditions at the facility that may influence the migration of any potential releases
7. The potential for exposure of humans and environmental receptors if releases were to occur from the unit

Once the decision is made to utilize a CAMU in a corrective action, and after the previously described supporting information has been developed, the TU will normally be approved under an EPA-initiated permit modification (§270.41) or according to the permit modification procedures of §270.41 for owner/operator-initiated modification. This will generally be the case for TUs that are part of a large-scale corrective action at a facility that requires agency approval through a Class III or agency-initiated permit modification. Justification for the TU, in the form of an evaluation against the seven criteria/standards, must be included in the permit modification. In addition, the permit modification will include the design, operating, and closure requirements for the TU.

In cases in which the TU is not part of a larger permit modification for a selected remedy (i.e., a unit to handle or store investigation-derived waste or remediation waste generated from remedial activities that do not require a Class III or agency-initiated permit modification), the owner/operator will be responsible for requesting approval of the TU in the form of a Class II permit modification (§270.42). If the operation of the TU is time critical (i.e., necessary to contain a release or to protect human health and the environment), the regional administrator may approve a 180-d temporary authorization for the unit upon request by the owner/operator according to the procedures under §270.42.

The rule allows for a 1-yr operating limit for a TU. EPA clearly states in the preamble to the rule that "a one-year time limit for temporary units is reasonable and appropriate." At the end of the time limit, the owner/operator will be required to cease management of remediation wastes in the TU and to initiate the closure requirements specified in the permit modification (§264.553(d). If continued operation of the TU is necessary or desirable for satisfactory completion of the selected remedy, the regional administrator has the authority to grant up to a 1-yr time extension beyond the time limit originally specified for the unit. Extensions will be limited to circumstances in which the owner/operator can prove that the extension is necessary to ensure timely and efficient implementation of remedial actions at the facility and that continued operation of the unit will not pose a threat to human health and environment.

7.6 *Review of natural attenuation remedy*

The macroengineering approach to groundwater cleanup includes as a

130   *Macroengineering: An environmental restoration management process*

at the point of exposure (or potential exposure) have been approved for a few sites, including the Pickettville Road Landfill (Jacksonville, FL).

Should natural attenuation be relied upon for attaining remedial goals, the site characterization program must sufficiently define site hydrogeologic conditions to support the passive groundwater remediation strategy inherent in the natural attenuation approach. Significant questions regarding the definition of the plume's characteristics (lateral extent, vertical extent, fate, and transport characteristics) must be answered. If one assumes institutional control is plausible and allows for a passive natural remediation process, EPA standards (40CFR192, Groundwater Standards for Remedial Actions at Inactive Uranium Processing Sites; Final Rule, for example) call for a demonstration that the natural remedy can meet groundwater standards within a 100-yr time frame. Specifically:

> ... for aquifers where compliance with the groundwater standards can be projected to occur naturally within a period of less than 100 years, and where the groundwater is not now used for a public water system and is not now projected to be so used within this period, this rule permits extension of the remedial period to that time, provided institutional control and an adequate verification plan which assures satisfaction of beneficial uses is established and maintained throughout this extended remedial period, ..."

> "(2)(I) If the Secretary determines that sole reliance on active remedial procedures is not appropriate and that cleanup of the groundwater can be more reasonably accomplished in full or in part through natural flushing, then the period for remedial procedures may be extended. such an extended period may extend to a term not to exceed 100 years if: (A) The concentration limits established under this subpart are projected to be satisfied at the end of this extended period, (B) Institutional control, having a high degree of permanence and which will effectively protect public health and the environment and satisfy beneficial uses of groundwater during the extended period and which will effectively protect public health and the environment and satisfy beneficial uses of groundwater during the extended period and which is enforceable by the administrative or judicial branches of government entities, is instituted and maintained, as part of the remedial action, at the processing site and what-

Thus, a site hydrogeologic investigation must be conducted that can adequately define: (1) variability of hydrogeologic regimes, (2) fate and transport characteristics, (3) background, and (4) plume extent. This investigation should be aimed toward allowing a sufficiently rigorous and sophisticated model to be developed to assess the ability of natural remediation progress to achieve remediation goals within the specified maximum regulatory extent of time (100 yr). Specifically, with respect to monitoring under a monitored natural attenuation remedy, OSWER Directive 9200.4 states (page 22) that the monitoring program should be designed to accomplish the following:

- Demonstrate that natural attenuation is occurring according to expectations
- Detect changes in environmental conditions (e.g., hydrogeologic, geochemical, microbiological, or other changes) that may reduce the efficacy of any of the natural attenuation processes
- Identify any potentially toxic and/or mobile transformation products
- Verify that the plumes are not expanding (either downgradient, laterally or vertically)
- Verify no unacceptable impact to downgradient receptors
- Detect new releases of contaminants to the environment that could impact the effectiveness of the natural attenuation remedy
- Demonstrate the efficacy of institutional controls that were put in place to protect potential receptors
- Verify attainment of remediation objectives

To address these issues, a groundwater-monitoring program needs to be comprehensive. Therefore, the items listed earlier should be explicit data quality objectives of the plume-monitoring plan. The passive nature of the remedial alternative for groundwater that essentially relies solely on dilution for contaminants, such as long-lived radionuclides, places an even greater importance on extensive groundwater monitoring. Unless a credible understanding of the groundwater flow and contaminant process can be developed, the true environmental impacts will remain unknown.

CERCLA regulations in 40 CFR 300.430(a)(iii) explicitly indicate that engineering controls (i.e., containment) and institutional controls (i.e., water use and deed restrictions) can be considered in the feasibility study as candidates for part of the total remedy for a contaminated site. These NCP provisions indicate that such controls may be appropriate where treatment is impractical. Related provisions in RCRA regulations at 40 CFR 264.94(b)

132   *Macroengineering: An environmental restoration management process*

Bibliography

1. Ecology, Washington State Department of Ecology, U.S. Department of Energy and U.S. Environmental Protection Agency, Letter to the Public (Attachment: Proposed Changes to the Hanford Tri-Party Agreement), May 20, 1991.
2. U.S. EPA, Federal Register Notice, 51 FR 4504, Washington, D.C., July 3, 1986.
3. U.S. EPA, CERCLA Compliance with Other Laws Manual, Draft Guidance, OSWER Directive 9234.1-01, Washington, D.C., August 8, 1988.
4. U.S. EPA, Federal Register Notice of Additions to the National Priority List, 54 FR 41015, Washington, D.C., November 3, 1989.
5. U.S. EPA, A Preamble to the Revised National Oil and Hazardous Substances Pollution Contingency Plan, 55 FR 8813, Washington, D.C., March 8, 1990.
6. U.S. EPA, Scorer's Notes, An RI/FS Costing Guide, Bringing in a Quality RI/FS on Time and Within Budget, EPA/540/G-90/002, Washington, D.C., February 1990.
7. Johnson, G.L. (U.S. EPA) and Wynn L.H. (Research Triangle Institute), A management systems review of the superfund RI/FS: opportunities for streamlining, *Journal of the Air and Waste Management Association*, Vol. 46, May 1990.
8. Myers, R.S. (Pacific Northwest Laboratory), The observational approach for site remediation at federal facilities, Proceedings of the Sixth Annual Waste Testing and Quality Assurance Symposium, Hanford, WA, July 1990.
9. OTA, U.S. Congress, Office of Technology Assessment, Complex Cleanup: The Environmental Legacy of Nuclear Weapons Production, OTA-O-484, Washington, D.C., February 1991.
10. U.S. EPA, Draft Final Report, Guidance Document for Closure and Post-Closure Costs, Vol. I, Unit Costs, EPA Contract Nos. 68-01-6621 and 68-01-6861, Washington, D.C., May 1986.
11. U.S. EPA, A Compendium of Superfund Field Operations Methods, EPA/540/P-97/001, Office of Emergency and Remedial Response, Section 8.4, Washington, D.C., December 1987.
12. U.S. EPA, Center for Environmental Research Information, Seminar Publication: Requirements for Hazardous Waste Landfill Design, Construction, and Closure, Washington, D.C., June–September 1988.
13. U.S. EPA, Groundwater Standards for Remedial Actions at Inactive Uranium Processing Sites; Final Rule, Washington, D.C., 1998.
14. U.S. EPA, OSWER Directive 9200.4-17P, Use of Monitored Natural Attenuation at Superfund, RCRA Corrective Action, and Underground Storage Tank Sites, Washington, D.C., April 1999.

chapter 8

Establishing cleanup objectives and natural resource damages

Although conceptual in nature, the macroengineering approach must provide sufficient detail to inform and define targets so that approaches and techniques, facilities, equipment, equipment modifications, and technology development opportunities can be identified and selected. The study must also serve as a physical conditions baseline, or benchmark, for relative comparison of subsequent and more detailed studies, and serve as a planning tool for decision makers and the public. To achieve a successful benchmark, the study must establish clear, consistent cleanup objectives. It is recognized that the cleanup levels may be subject to future debate and modification as the approach is further developed or new information becomes available. Although the use of "proposed cleanup levels" does not imply that they will be the ultimate "enforceable cleanup levels," it does provide a benchmark to assess cleanup-level cost impacts as the site embarks on negotiation with the regulators.

8.1 Macroengineering cleanups

Action levels (ALs) are the health and environmentally based maximum concentration of the contaminant of concern (COC) to be present at the termination of cleanup. ALs must be reviewed and certified by the administrative authority to be indicators for the protection of human health and the environment. ALs are defined in the preamble to the proposed Subpart S regulations and are also proposed in 40 CFR Part 264.521. The latter are established as sensitive "trigger" levels for the corrective measures study.

In this context, ALs are developed with the presumption of potential

134   *Macroengineering: An environmental restoration management process*

for 40 CFR Part 264.521 are proposed, not promulgated, under Subpart S (with the exception of maximum concentration levels [MCLs]) for certain highly toxic compounds for which the risks. ALs, in a similar sense to the MCLs, are specified in permits and orders. ALs may also serve as "target levels" or points of departure for setting cleanup standards.

When promulgated standards such as MCLs are available, they can be used as ALs, unless the "permittee" or potential responsible party (PRP) wishes to make the determination on a more specific basis using site parameters and specific laboratory tests. When promulgated standards are not available, ALs will typically be based on reference doses (RFDs) or carcinogenic slope factors (CSFs) published by USEPA or one of the health agencies. The latter is based on 1×10^6 risk for Class A and Class B carcinogens and 1×10^5 risk for Class C carcinogens.

These calculated ALs are available in Appendix A of the Subpart S preamble and represent one excess death in the universe of 70-year-old lifetimes per 1,000,000 citizens and for one excess death in a population of 100,000 persons. Class A and Class B carcinogens are known initiators of cancer, whereas Class C compounds are believed to be promoters of cancer.

For groundwater, the AL is the MCL if available. If an MCL is unavailable, the AL is calculated using RFDs and CSFs. The exposure assumption is an adult weighing 70 kg and drinking 2 l/d for 70 years. The point of measurement is any point in the plume.

For inhalation of contaminated air, the threat is normally considered the result of vapors emanating from a contaminated site. The AL is calculated using RFDs and CSFs with the exposure assumption of an adult weighing 70 kg, inhaling 20 m^3/d for 70 years. The point of measurement is taken to be the maximum contamination along the outermost facility boundary.

For surface water, the AL may be based on: (1) the numerical State water quality standards (Clean Water Act) when available; (2) numeric interpretations of narrative State standards if established; (3) MCLs, if the water body is designated for drinking water and an MCL is available for the specific constituent; and (4) may be calculated using RFDs and CSFs for drinking-water suppliers. The exposure assumption (for drinking water) is a 70-kg adult drinking 2 l/d for 70 years. Here, the point of measurement is the point at which release enters the surface water (although one could argue that the nearest point at which surface water is used as a drinking water source is more applicable).

Soil ALs are based on calculations using RFDs and CSFs. Exposure assumptions differ for carcinogenic vs. noncarcinogenic cases. For carcinogenic cases, the exposure assumption is a 70-kg adult ingesting 100 mg/d for 70 years. For noncarcinogenic cases, the exposure assumptions is a 16-kg

Chapter 8: Establishing cleanup objectives   135

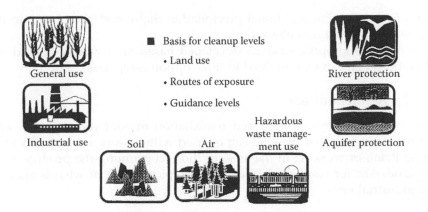

General use

Industrial use

■ Basis for cleanup levels
  • Land use
  • Routes of exposure
  • Guidance levels

Soil   Air

Hazardous
waste manage-
ment use

River protection

Aquifer protection

Figure 8.1 Consistent definition of point of compliance and background critical.

accessibility of the site. An element of practical realism benefits all. In general, cleanup objectives should reflect the intended or more probable future use of the land or groundwater under remediation, routes of exposures, and guidance levels. Target cleanup levels for lands designated for residential/ agricultural use may differ from levels for lands designated for industrial use or, for that matter, special use such as hazardous waste management. For this study, the term "general use" refers to potential rural/residential/ agricultural use and includes areas designated as wildlife areas, wetlands, or similar uses.

As shown in Figure 8.1, cleanup and restoration can typically be evaluated based on three cleanup options (general use vs. industrial use vs. hazardous waste management use). Each option reflects a different future land use, thereby providing a means to evaluate the impact of cleanup levels on costs and schedules. Similarly, groundwater cleanup should be evaluated and compared based on multiple projected groundwater uses to assess the cost/ benefit relationship of groundwater cleanup levels. Also, soil/source unit closures and groundwater closures require consistent definition of point of compliance and background. Typical cleanup options and scenarios are summarized in the following.

8.1.1   Closure of soil/source units objectives

Clean closure offers an option to cease operation of the solid waste man-mass from receiving additional precipitation that could percolate into the waste and encourage mobilization.

Soil cleanup options can be developed based on a variety of land-use alterations that can be described in three option categories.

8.1.2   Residential use

This option requires cleanup and remediation of contaminated soils and source areas such that the land can be used without any adverse effects on those living or working in the area, or those consuming the produce from the land. ALs for residential land use are typically half of what is allowed for industrial sites.

8.1.3   Industrial use

Under this option, the soils will be remediated to such a degree that those employed in this area will not be adversely affected by previous operations in this area. The land will not be restored sufficiently to support residential use. Industrial sites are often treated by at least partial removal of wastes to depths at which volatized components are either collected or managed or for which large outdoor areas are paved with asphaltic concrete, sometimes with vapor barriers.

8.1.4   Contained management use

Contained management areas (CMAs) should typically be located centrally on a given site, away from any major exposure pathway. Typically, past and present hazardous waste management activities in these areas would be extensive and would have generated a significant degree of contamination that makes clean closure cost prohibitive. Because of the degree of existing contamination in this vicinity, the area's best practical land use is as a central disposal facility for all wastes retrieved from the other areas of the site, consistent with a contained management philosophy. As a result, the cleanup option for this contained management area should be less stringent than the general use and industrial use option cited for clean closure.

Recognize that site location is paramount in the decision to opt for a contained management use strategy. Urban areas are not ideal for contained management use strategy for highly charged waste management circumstances, such as radioactive waste management, be it mixed or low level. In such circumstances, if the contained management use strategy is selected, more rigorous design concepts and a stronger public outreach program must

Chapter 8: Establishing cleanup objectives 137

The approaches discussed here address groundwater condition exclusively as the major decision factors in facilitating large-scale remediation and to avoid potential problems from overlap between the specific SWMU areas. Groundwater represents the principal means of potential communication in the event of failure of the final cover or containment liners. Three scenarios were selected to approximate future groundwater use. The scenarios and corresponding groundwater objectives are described in the following.

8.1.5.1 *Option 1: Large-scale surface water body protection (rivers, lakes, bays, etc.)*

Groundwater contaminant discharges to large surface water bodies should be reduced to acceptable concentrations within a defined period of time (i.e., 25 years). Site groundwater should not be restored or available for general use. Institutional controls on groundwater extraction or infiltration should be imposed and remain in perpetuity. For example, groundwater in the contained management areas typically would not be available for general use, but would have long-term (perpetuity?) institutional controls imposed. However, proving the viability of long-term (i.e., perpetuity) controls is oftentimes difficult, if not unfeasible.

8.1.5.2 *Option 2: Long-term groundwater control and restoration*

Groundwater contaminant discharges to the nearest major surface water body should be reduced to acceptable concentrations within a defined period of time (i.e., 25 years). The long-term good is for groundwater in the areas in which soil/source removal occurs (i.e., clean closure) to be restored to acceptable contaminant concentrations for general or industrial use. Institutional controls would be imposed until groundwater is restored. Currently, regulations call for a demonstration that groundwater cleanup can be achieved within 100 years where natural attenuation mechanisms are being called upon to achieve cleanup. After aquifer restoration, general groundwater use in the restored areas would be subject to the appropriation doctrine of the State and water right permit acquisition.

8.1.5.3 *Option 3: Groundwater cleanup within a given time frame (i.e., 25 years)*

Groundwater contaminant discharges to the nearest major surface water body would be reduced to acceptable concentrations, under stipulations of the National Pollutant Discharge Elimination System (NPDES), as

138   *Macroengineering: An environmental restoration management process*

Table 8.1 Applicable EPA Guidance for Dose and Risk Assessment

"Risk Assessment Guidance for Superfund," *Human Health Evaluation Manual (Part A)*, Volume 1 (DPA/540/1-89/002).

"Role of the Baseline Risk Assessment in Superfund Remedy Decisions," OSWER Directive 9355.0-30.

"Supplemental Guidance: Standard Default Exposure Factors," *Human Health Evaluation Manual*, OSWER Directive 9285.6-03.

"Risk Assessments Methodology, Environmental Impact Statement," *NESHAPS for Radionuclides, Background Information Document*, Volume 1, EPA/520/1-89-005 (September 1989).

"Limiting Values of Radionuclide Intake and Air Concentration and Dose. Conversion Factors for Inhalation, Submersion, and Ingestion," Federal Guidance No. 11, EPA 520/1-88-020 (September 1968).

8.1.6   Target levels

In the U.S., the selected cleanup objective standards should be developed based on EPA- and State-mandated standards. Applicable EPA guidance for dose and risk management is continually undergoing revision but applicable guidances are presented in Table 8.1. The strategy for setting these levels should be based on the following:

- If State or EPA reference levels exist, the most conservative available value was used.
- If no reference dose or level is available for a compound, a level should be determined based on the toxicity of the compound to ecological organisms.
- If the compound is judged to be relatively nontoxic or to have a high background in the soil, background levels should be referenced.
- In the case of radionuclide levels, these should be set using NRC-mandated levels for soils and a 4-mrem annual exposure dose for groundwater.

Special consideration should be given when establishing cleanup levels for groundwater that may discharge to a major surface water body. Consideration should also be given to the potential impact that groundwater may have on the present ecology of the major surface water body. To that end, examples of recommendations for chronic aquatic levels of chemical constituents are provided by the references cited in Table 8.1. Soil cleanup levels should be selected taking these levels into consideration.

Table 8.2 Macroengineering Cleanup Level Tables — Chemical

Contaminants	Drinking water mg/L	Chromic aquatic mg/L	Ground-water mg/L	Soil (residential) mg/kg	Soil (industrial) mg/kg
Gross alpha[m]	15 pCi/L[a]		15 pCi/L[d]		
Gross beta[m]	50 pCi/L[a]		4 mrem/year[d]		
pH	6.5–8.5[b]	6.5–8.5[l]	6.5–8.5[g]	6–9[g]	
Total coliform	>10% tests[a]	or 100	org/100 ml	Sample	
Total organic carbon	1.0				
Total organic halogen	0.32				
Aluminum		0.87[i]	5[i]	10,000[g]	10,000[p]
Antimony		1.6[l]	0.146[i]	30[l]	1,313[j]
Arsenic[m]	0.05[b]	0.048[l]	0.005[d]	20.0[d]	200[d]
Barium	1[a]	1[e]	1.0[d]	100[cg]	3,500[g]
Beryllium[m]		0.0053[l]	0.005[d]	0.2[j]	20[k]
Cadmium[m]	0.01[a]	0.0011[l]	0.005[d]	2.0[d]	10[d]
Calcium			<500[g]	14,000[g]	400,000[g]
Copper	1[b]	0.012[l]	1[d]	100[d]	10,000[p]
Iron	0.3[b]	1[l]	10	25,000[c]	100,000[g]
Lead[m]	0.05[a]	0.0032[l]	.005[d]	250[d]	1,000[d]
Magnesium			<400[g]	5,000[c]	5,000[g]
Manganese	.05[b]		<400[g]	3,000[c]	4,000[g]
Mercury	0.002[a]	0.000012[l]	0.002[d]	1.0[d]	1.0[d]
Nickel[m]		0.160[l]	0.7[k]	2,000[j]	10,000[p]
Potassium			5[g]	2,000[c]	30,000[g]
Selenium	0.01[a]	0.035[l]	0.01[d]	240[j]	10,000[p]
Silver	0.05[a]	0.00012[l]	0.05[d]	200[d]	1,000[n]
Sodium			100[g]	1,000[c]	7,500[g]
Strontium	8 pCi/L[a]			600 pCi/L	
Thallium	.013[d]	.013[i]	0.0002[j]	6.4[j]	280[j]
Vanadium			.02[j]	720[j]	3,600[n]
Zinc	5[b]	0.110[l]	.48[j]	16,800[j]	10,000[p]
Ammonium		.05[h]	.1[h]	2[c]	2[c]
Chloride	250[b]		<1,000[g]	10[g]	100[g]
Fluoride	4[a]		4[l]	400[ko]	400[g]
Nitrate	10[a]		20[l]	2,000[ko]	2,000[g]
Nitrite			20[f]	2,000[ko]	2,000[g]
Sulfate	250[b]		<2,000[g]	30[cg]	10,000[g]
Phosphate			<1,000[g]	2,000[cg]	5,000[g]
Arochlor 1260[m]		0.00002[l]	0.0001		
Arochlor 1248[m]		.00002[l]	0.0001		
Chloroform[m]	0.10[d]	1.2[l]	0.023[j]	100[j]	4,375[j]
Dichloroethene[m]	0.007[a]	1.2[l]	0.020[j]	10[j]	100[k]
Methanol		100[e]	1.142[j]	114[ko]	570[kn]
Methyl isobutyl		9[h]	0.114[j]	4,000[j]	10,000[p]

Table 8.2 Macroengineering Cleanup Level Tables — Chemical (Continued)

Contaminants	Drinking water mg/L	Chromic aquatic mg/L	Ground-water mg/L	Soil (residential) mg/kg	Soil (industrial) mg/kg
Benzene[m]	0.005[a]	0.053[l]	0.005[d]	0.5[d]	0.5[d]
Ethyl benzene[m]		3.2[l]	0.03[d]	20.0[d]	20[d]
Total xylenes		0.36[l]	0.02[d]	20.0[d]	20[d]
Toluene		1.75[l]	0.04[d]	40.0[d]	40[d]
Acetone			.22[j]	8,000[j]	10,000[p]
Boron			.21[j]	7,200[j]	10,000[p]
Bis-2-ethyl hexyl Phthalate[m]		3[l]	.0009[j]	50[k]	5,000[k]
Chromium (IV)[m]	.05[a]		0.05[d]	80[d]	500[d]
Chlorobenzene		50[l]	0.0003[j]	1,600[j]	8,000[n]
Cyclotetrasilo-xane octomethyl			1[e]	10[e]	10[e]
Cyanide		5[l]	0.0003[j]	1,600[j]	8,000[n]
Diesel fuel				200[d]	200[d]
Hexane			100[h]	10,000[ko]	10,000[p]
Hydrazine[m]			4×10^4	0.3[j]	45[j]
Herbicides[m]			.010[l]	1[ko]	1
Lillium			70[g]	200[g]	200[g]
Morpholine		100[e]	7 (1/10)	70[h]	350[kn]
4 Methyl 2 pentanone (methyl iso-butyl ketone)		9[h]	0.114[j]	4,000[j]	10,000[p]
Oxalate			20[h]	2,000[ko]	10,000[kn]
Sulfamate			2,000[f]	2,000[f]	10,000[f]
Tetraethylpyro-phosphate			0.001[j]	40[j]	1,750[j]
Tetrahydrofuran			0.5[h]	50[ko]	250[kn]
Thiourea[m]			5×10^4	0.5[l]	65[j]
VOCs	0.1[a]				

Note: a = Primary drinking water standards; b = Secondary drinking water standards; c = Background; d = State of Washington; e = Dangerous properties of industrial materials, Sax; f = By comparison; g = Soil chemistry of hazardous materials, Dragum; h = Handbook of environmental data on organic chemicals, Verschueren; i = Toxicology profiles; agency for toxic substance and disease registry; j = By ecology MTCA formulae; k = Proposed AL; l = EPA reference dose; m = Carcinogen; n = 5 times the residential soil level; o = 100 times the groundwater level; and p = 1% upper limit.

Table 8.3 Radionuclide Macroengineering Cleanup Limit Tables

Contaminants	Drinking water pCi/L	Chromic aquatic PCi/L	Groundwater PCi/L	Soil (residential) pCI/gm[b]	Soil (industrial) pCi/gm[c]
^{60}Co			200	1	5,000
^{98}Tc			4,000	1,750	100,000
^{147}Pm			4×10^4	1,700	20,000
^{235}U	20		24	15	15
^{238}U	20		24	50	50
Tritium	20,000		80,000	35,000	1×10^7
^{137}Cs	10		120	3	20,000
^{90}Sr	8		40	13	600
^{241}Am			1.2	20	20
^{242}Am			1.2	20	20
^{243}Am			1.2	20	20
^{238}Pu			1.6	75	75
^{235}Pu			1.2	75	75
^{240}Pu			1.2	75	75
^{106}Ru	30		240	15	15
^{129}I	1		20	50	4,000
^{241}Pu			80	2,500	2,500
^{226}Ra			3^d	5	60
^{152}Eu			8×10^2	3	3,000
^{154}Eu			8×10^2	3	3,000
^{155}Eu			4×10^3	100	20,000
^{161}Sm			16×10^3	500 H	2,000 H
^{134}Cs			80	2	10,000
^{125}Sb			2×10^3	5	60,000
^{113}Cd			32	1 H	5 H
^{103}Ru			2×10^3	60 H	250 H
^{107}Pd			4×10^4	1,250 H	5,000 H
^{94}Nb			12×10^2	35 H	150 H
^{93}Zr			36×10^2	125 H	500 H
^{63}Ni			12×10^3	3,900	100,000
^{79}Se			8×10^2	25 H	100 H
^{14}C			7×10^3	870	3×10^7
^{41}Ca			4×10^3	125 H	500 H
^{51}Cr			4×10^4	10 H	50 H

Note: a = 0.04 of Derived concentration guide for public exposure approximate 4 mrem exposure; b = WHC-CM-7-5, Table K-1, unless otherwise marked; c = WHC-CM-7-5, Table K-2, unless otherwise marked; d = State of Washington; H = Derived concentration guide for public exposure approximate 25 mrem annual exposure; and H = Derived concentration guide for public exposure approximate 100 mrem annual exposure.

142   *Macroengineering: An environmental restoration management process*

Selection of the cleanup criterion for contaminated sediments also poses significant overall site implications. Contamination of sediments is widespread throughout U.S. coastal waters and especially its harbors. Such contamination should be expected at virtually all aquatic sites in the U.S. that lie in the watershed of urban and/or industrial development. Sediment contamination represents a potential key pathway of long-term contamination to the food chain. In particular, it may impair the marketability of seafood and may also interfere with improvements in water quality to the extent that it could cause chronic damage to aquatic biota and ecological systems. Since the passage of NEPA in 1969, sediment contamination has imposed a severe impediment to maintenance of the dredged channels of our navigable rivers. Even aside from the disturbance generated by dredging, there is the significant problem of disposal of the contaminated soil. These are unique problems of coastal site remediation. Although the problem is well recognized and potentially far reaching, its extent has been poorly quantified. Extensive studies have been published on this subject by the U.S. Army Corps of Engineers waterways experiment station. The leading areas of problems faced the Hudson River below Schenectady, NY (NY DEC state lead) and New Bedford Harbor, MA (federal EPA lead), both of which suffer from heavy concentrations of PCBs from their upgradient electrical utility industries.

Potential impacts on the regulated community are dramatic. Specific areas of concern include:

- More stringent discharge limits
- Greater cost and difficulty of dredging navigable waterways
- Increased Superfund (and RCRA) cleanup costs
- Costly natural resource damage claims
- Adverse impacts on fishery harvest and marketability
- Increased testing and monitoring costs

For example, Table 8.4 presents representative guideline numbers for sediments with low organic carbon content; sediment quality consideration (SQC) values can be quite low.

All approaches to the development of SQCs have strengths and weaknesses. Relative positive SQC factors are:

- Accounts for site-specific factors
- Applicable nationwide
- Based on well-established water quality criteria (WQC)
- Accounts for bioavailability

Chapter 8:   Establishing cleanup objectives　　　　　　　　　　　143

Table 8.4 Representative Guideline Numbers for Sediments with Low Organic Carbon Content (Sediment Criterion Values Can Be Quite Low)

Pollutant	Interim or preliminary sediment qualification criterion	AL for soils	FDA AL in fish	Adjusted NS&T" "high" concentration	Possible biological effect concentration
PCBs	FW: 19:g/gC (mean) SW: 33 :g/gC (mean)	Residential: 1.0 :g/g Industrial: 10.25 :g/g (.09 :g/g)	2.0 :g/g	0.2 :g/g (dry)	0.4 :g/g
t-DDT		2 :g/g	5.0 :g/g	0.04 :g/g	0.35 :g/g
t-PAH				3.9 :g/g	35 :g/g
Acenaphthene	FW: 138 :g/gC (mean) SW: 243 :g/gC (mean)				
Toxaphene		0.6 :g/g	5.0 :g/g	—	—
Mirex			0.1 :g/g	—	—
Heptachlor/ H-Epox		0.8 :g/g	0.3 :g/g	—	—
Dieldrin	FW: 20 :g/gC (mean)	0.4 :g/g	0.3 :g/g	—	—
Endrin	FW: 4.03 :g/gC (mean) SW: 0.73 :g/gC (mean)	20 :g/g	0.3 :g/g	—	—
t-Cdane		0.5 :g/g	0.3 :g/g	0.055* :g/g	0.006 :g/g
Cd		40 :g/g		280 :g/g	270 :g/g
Pb		—		87 :g/g	110 :g/g
Hg		20 :g/g		0.51 :g/g	1.3 :g/g
Ag		200 :g/g		1.2 :g/g	2.2 :g/g

- Ignores impacts on biota of direct contact and ingestion
- Assumes chemical equilibrium
- Cannot deal with mixtures
- Limited to chemicals covered be WQC and with partition coefficients
- Makes dubious assumption of constant BCF for various contaminants and organisms
- Most readily applicable to hydrophobic neutral organics

300 million cubic yards (mcy) of dredged sediments annually. Special and occasional operational needs lead to an average of permits covering another 100 to 150 mcy. Adoption of SQCs as state water quality standards or as Superfund ARARs could adversely affect the COE dredging program.

Currently, only a limited amount (3 to 10 mcy) of dredged sediments are deemed "contaminated" and unsuitable for unconfined open water disposal, but this is because COE utilizes "effects-based" bioassay-testing criteria at present to model the impact on riverine or marine life.

As many as 13% of Superfund NPL sites may involve contaminated marine sediments (NRC, 1989). The implication of SQCs for Superfund sites is that the number of contaminated sediment sites and stringency of required cleanup at these sites are likely to increase under pressure from environmental groups.

§/105 of SARA requires the hazardous ranking score (HRC) to take into account damage to natural resources that may affect the human food chain. Revised HRS promulgated on December 14, 1990 (55 FR 51531), specifically identifies food chain contamination as a subpathway of surface water pathway. Thus, this clearly positions SQCs as a potential ARAR for use in setting cleanup standards and for defining cleanup "ALs" (e.g., 8/90 Superfund PCB guidance). As a result, SQCs may be used in natural resource damage claims in defining "injury *per se.*" Whereas EPA and its legal representative, the U.S. DOJ, are reticent to cite natural resource damage claims, such claims are employed mainly when the PRP is noncooperative or legal action is enjoined by environmental activist and watchdog groups.

Natural resource claims have a rising profile and have been supported by a National Academy Marine Board study (NRC, 1989) that found contaminated sediments to be widespread throughout U.S. coastal waters and have potentially far-reaching environmental and health significance. The study identified contaminated sediments as a cause for concern in terms of contamination of seafood and potential impairments of mammalian reproduction. However, the study also pointed out that till date the issue has been inadequately studied to pinpoint or prioritize candidates for remediation action, no lead agency is taking responsibility, and no generally accepted AL or cleanup standard is in place. Thus, the NRC study confirmed that contaminated sediments are a significant problem requiring more systematic study and regulatory attention.

Particular insights concerning sediment cleanup technology and policies that emerged from the study are that in-place capping can be a useful isolation and contaminant option (but not a "preferred treatment" under SARA §121(b)?). Currently, cleanup decisions for navigable rivers and harbors could be expedited by relying on Clean Water Act (§115/404) rather than CERCLA/SARA authorities. Use of the Clean Water Act (CWA) for

Chapter 8:   Establishing cleanup objectives   145

haphazard ways in which "how clean is clean" standards are set. In nearly all cases, this situation points to the need for a carefully and appropriately devised scenario for selection of ALs in which word walks are used to define and portray reasonable receptors and the pathways by which they come into contact with the sediment contamination. Scenarios are useful in identifying and distancing potential concerns that do not survive the scrutiny of detail explanation.

Other insights concerning sediment cleanup technology and policies that came out of the study include the fact that no-action may be the acceptable and/or inescapable solution if contaminants degrade or are buried rapidly. Furthermore, downstream sediment contamination tends to involve larger volumes of less-contaminated material than at upland hazardous waste sites. High-tech remedial technologies are usually cost effective only in small, highly contaminated sites (high total ppm).

Sediment cleanup policies have tended to lag behind the technology; however, to be cost effective, cleanup must often be pursued within relatively narrow time windows.

Expedited response (interim remedial measures [IRMs]) should be instituted where the potential for significant contamination spread is high. Furthermore, a tiered strategy should be considered to permit rapid initial response action where contaminants are being spread around, and pursue opportunities to "piggyback" on dredging technologies, strategies, and current projects in the site area.

Creative sampling strategies can sometimes dramatically reduce costs. Typically, sediment contamination may be present over large distances downstream of the discharge points. Be aware that different sampling designs will give dramatically different indications of contamination status. It is easy to spend hundreds of thousands of dollars on sampling and analysis in lengthy waterway segments using conventional grid patterns. Some ways to keep costs down include: understanding the hydrogeology; sample in phases; first characterizing by grain size and type of contamination (TOC); conducting stratified random sampling (focusing on sediments types likely to attract contaminants); treating data as spatially dependent; and compositing and archiving samples, as appropriate.

Oftentimes, "piggybacking" of sediment cleanups with navigational dredging projects can save time and money. It is taken into account that Superfund remediation projects can take a decade or more to run their course, whereas, dredging projects can sometimes proceed more routinely and expeditiously. Currently, CWA §115 and 1990 WRDA provide additional authority for removal of contaminated sediments. So, if sediments need to be dug up and removed, using dredging authorities rather than Superfund

146   *Macroengineering: An environmental restoration management process*

The decision to manage in place, or to remove and relocate sediments requires consideration of the trade-off relative to the additional risk potential for contaminant mobilization and environmental release that accompany excavation.

Capping (final cover) is not viable if sediments are in a navigation channel or in an area of groundwater upwelling. Also, capping or other remedial measures must be coupled with or preceded by action to control the sources of contamination. In some instances, capping may be combined with an *in situ* treatment, such as bioremediation. However, capping may not qualify as a "preferred treatment" approach under SARA §121(b).

As noted in the opening introduction, the site should take claimed impacts on drinking water with a grain of salt. In particular, be suspicious of claimed drinking water impacts of hydrophobic sediments contaminants. Sediment contaminants tend to be either heavy metals or dense nonaqueous phase liquids (DNAPLs) with relatively low water solubilities. When present in drinking water, sediment contaminants are usually in conjunction with fine particles or an oil sheen. Thus, it is inappropriate to apply drinking water standards under such circumstances. HRS scoring of such sites should be closely reviewed to ensure that NPL listing was not based on overstated drinking water risks.

Fishing bans or advisories based on claimed health hazards should also be carefully reviewed. Again, health risk claims based on food chain contamination should not be swallowed too readily. Health risk assessments of the threat due to consumption of contaminated fish can be greatly exaggerated based on inflated assumptions concerning:

- Dietary intakes of contaminated vs. uncontaminated species
- Constant high-level exposures to the same fish throughout a 70-year life expectancy
- Choice of high-to-low-dose extrapolation models for carcinogens (and nonthreshold assumption)
- Artificialities in computation of CPFs
- Errors/variability in sampling techniques

Many health-based risk assessments are formulated on an inaccurate geologic assessment of the actual framework and natural mineralogical nature of the host sediment. Once unchallenged fishing bans are in place, sociopolitical pressures may bring about large natural resource claims; hence, a proactive approach is warranted.

The site should also not ignore the "natural restoration" alternative. Natural restoration may be acceptable or advisable if the contamination rapidly degrades, or it is rapidly buried by natural deposition of clean sediments. Also, natural restoration may be advisable if artificial intervention

when "relevant and appropriate" under the circumstances of the release, they may not be "appropriate" when cleanup to meet the SQC could cause short-term habitat or benthos destruction or contaminant resuspension that outweighs long-term benefits. Compliance to such a standard is not technically feasible or would be cost-prohibitive; and application of another (less costly) standard yields equal or greater protection. Also, SQCs arguably do not have the status of enforceable requirements for ARAR purposes until they are adopted (and applied) as state standards.

8.2   Natural resource damage assessments

The stakes relative to national resource damage (NRD) claims are extremely high. To start with, from an NRD perspective, the cleanup standard is full restoration. This in itself is at odds with practical reality at most large-scale macroengineering-type sites. Furthermore, NRD issues typically begin where Superfund remediation leaves off. Although oftentimes NRD claims may involve impaired reproduction of numerous fish, bird, and mammal species, the definition of protected "natural resources" is extremely broad and could involve contamination of very large but inaccessible areas typical in macroengineering-type settings. For example, compensable injuries may include the public's lost "psychic" enjoyment of resources that they were unlikely to ever use.

The other frustrating aspect to NRD issues is that the targets are unpredictable. Although only designated "trustee" agencies can initiate NRD claims, federal trustees are becoming increasingly but sporadically active and some states (e.g., CA, NY, NJ, OH, MI, CO, etc.) are becoming more active under both federal and parallel state law. In addition, Indian tribes are starting to get involved, and, in some instances, are the dominant interest parties. It should be noted that NRD targets are not limited to designated Superfund sites. NRD claims can have devastating economic impacts because "cleanup" liability is based on restoration without regard to risks, costs, or benefits. Damage payment by a single party at an individual NRD site can approach or exceed $100 million.

Natural resource damage claims can be brought by any trustee of these resources against PRPs owing to prior hazardous substance releases. For the trustees to prevail on an NRD claim, all three elements identified in Figure 8.1 must be demonstrated.

Figure 8.2 depicts the NRDA process that a trustee must follow. Any trustee can trigger the process, and the trustee receives rebuttable presumption

"Injury above thresholds"

148   *Macroengineering: An environmental restoration management process*

(e.g., the PRP must prove to the court that this process is inappropriate for the types of damages being alleged). As seen in this figure, the NRDA process includes preassessment, development of a damage assessment plan, the performance of damage assessment studies for each injured resource, and postassessment activities.

The process results in many steps and can require dozens of separate studies, reports, and public comment periods unless an early settlement can be reached.

Figure 8.3 provides two prototypical examples of the impact of the NRDA expansion. The stakes are particularly high where potential contamination of very large areas and potential impaired reproduction of numerous fish, bird, and mammal species may have occurred.

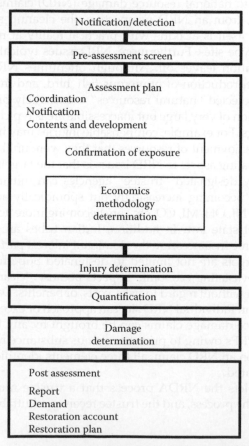

- The concept of non-use economic benefits is still evolving, but includes such components as bequest and experience values that add to a PRP's liability. The bequest value is the value associated with the knowledge that a resource is preserved for future generations. The existence value is the value associated with the knowledge that a resource is preserved, regardless of its use-related benefit to a person. For example, one may never visit the Everglades, but still attach a value to its existence in a healthy state (existence value) and to the knowledge that it will be around for one's children and grandchildren to enjoy (bequest value).

- The trustee can include any agency which "appertains to" or otherwise "controls" the resource, or any quasi-government organization with jurisdiction over some aspect of a resource. For example, Canada could conceivably file a case based on the loss of waterfowl which migrate to Canada. Likewise, a Native American tribe denied access to migratory salmon could initiate a case.

Figure 8.4 Prototypical example of NRD claim expansion impact.

Companies associated with long-term aquatic sediment contamination may be most vulnerable to NRD claims. Thus, it is critical for large-scale environmental restoration sites to take natural resource damage potential into account during the baseline engineering study development.

Figure 8.4 defines the type of macroengineering environmental restoration sites that should worry about NRD claims. It should be noted that most NRD cases to date have involved contamination of aquatic sediments by "notorious" toxic pollutants.

In effect, a site can follow two strategies: (1) lie back and wait or (2) be proactive. The advantages of a proactive approach are that the advance planning keeps transaction costs down and reduces ultimate NRD liability exposure. The proactive strategy allows a site to identify potential contributors to future NRD actions and put programs in place to curtail the environmental restoration program's impact on the community as a whole and mitigate grounds for the NRD's action. It should also be noted that the recalcitrants who "lie back and wait" make more attractive targets.

150   *Macroengineering: An environmental restoration management process*

appropriate legal releases are obtained. In particular, the site should consider cost-effective mitigation measures to minimize the impact of environmental restoration activities. In short, make sure you do not make the problem worse. In particular, characterization and remediation efforts should also include gathering evidence in support of potential NRD "defenses" (e.g., pre-1980 release; federally permitted release). Other proactive actions include beginning to assemble exculpatory records and identifying other potential NRD contributors.

The most effective way to reduce transaction costs and the size of settlements or awards in an NRD case is to keep the scope of the damage assessment plan (DAP) and the IDP within proper bounds. There are ten technical defense tips to mitigate NRD claims. Focus early and often on the following likely "weak links" in the trustees' case:

If the site is a potential target of an NRD claim, a *site natural resource damage assessment (NRDA) model* should be developed to assess whether the NRD issues are appropriately addressed during the remediation phase at the particular site through cost-effective mitigation measures.

As stated earlier, the concept of what constitutes natural resource subject to damages and the nature of those damages has expanded. The PRPs are now potentially liable for both residual damages and restoration (or replacement) costs, unless the they can prove that the restoration costs are "grossly disproportionate" (Ohio vs. U.S. DOI). Unfortunately, there are no clear definitions of *feasibility* or *grossly disproportionate*, making it imperative that the potential PRPs be proactive in defining these terms relative to their case.

The site NRDA model provides the basis for a site to effectively gather evidence in support of potential NRD "defenses" (e.g., pre-1980 release; federally permitted release). The successful site management will focus early on the areas in which potential "unrealistic expectations" in the trustees' case may be. Figure 8.5 identifies the ten areas of "unrealistic expectations" that the model looks for in an NRD claim. The list also provides a set of criterion to use to effectively reduce the transaction costs and potential size of settlements or awards in cases in which claims may be valid, but expectations are unrealistic. Again, the critical issue in the latter circumstance is to keep the scope of the DAP and the IDP within proper bounds to ensure that effective cleanup and improvement is not delayed through legal brain lock.

Different strategies are available for dealing with and mitigating NRD claims. A combination of approaches can be used to resolve NRD problems, but most of these strategies are associated with the timing or phasing of resource restoration or enhancement initiatives as part of cleanup operations. The following provides a discussion of alternate approaches to NRD issues at different stages of cleanup.

Chapter 8: Establishing cleanup objectives   *151*

The types of environmental restoration sites that should worry about NRD Claims:

- Are on a governmental hazardous substance list (e.g., CERCLIS, TRI database, NPL, CERCLA/SARA notification)

- Have been operating for 20+ years

- Have released large (over time) quantities of persistent, toxic, and bioaccumulative substances — especially if "notorious" (PCBs, DDT, dioxin, Pb, Hg, Cd, As)

- Are adjacent to "valuable" waterways, wetlands, waterfowl/wildllife refuges, endangered species habitat, tribal lands

- Are located in states with active NRD programs

- Don't have Federal permits that fully and specially authorize the hazardous substance releases in question

- Have contributed to extensive sediment contamination

> **It should be noted that most NRD cases to date have involved contamination of aquatic sediments by "notorious" toxic pollutants.**

Figure 8.5 The types of environmental restoration sites.

the off-site protection of threatened resources (e.g., purchase of some critical land for a new refuge or park). Low-cost, off-site mitigation activities, if adopted early, can greatly reduce a PRP's liabilities. For example, the placing of a relatively inexpensive salmon hatchery upstream or downstream of a potentially contaminated area near a

1. Size of the area allegedly impacted as a result of the claimed releases. Size of the study area was part of the study area polluted/affected even prior to the release "baseline"? Keep the targeted area as small as possible relative to the overall size of the facility.

2. Quantifiable and compensable injuries. Eliminate any "injuries" that cannot be translated into quantifiable lost uses or "services." Even if you can quantify the number of common and abundant fish affected, compensable injury must be measured on an incremental basis. Incremental losses to an abundant resource may be small.

3. Components of the injury. The key here is to make sure you are not blamed for injuries due to other causes. Segment cause ane effect relationships based on different contributory pollutants, different congeners, and other chemical, physical and biological factors.

4. Uniqueness of the resource. Injury to rare and unique, rather than common and abundant, species. The key here is that claims based on injuries to common and abundant natural resources should be measured at the margin.

5. Moving "targets." Can injury incurred by migratory or highly mobile birds and mammals that be potentially attributed to other geographically-specific release source(s).

6. Evidence of actual harm. The practicality of the links between available lab and field data and evidence of biological harm (to individuals, a population, or the species). Under what circumstances is it permissible to extrapolate from one species to another or to select one species as an indicator or surrogate?

7. Status of population. Assess evidence of population impacts. Be skeptical of damage claims for species that are thriving in the wild.

8. Scope of damages. Assess he relevance of pre-release "baseline(s)," rough mass-balance calculations, nearby control reference areas, and applicable exclusions from liability.

9. "Per Se" injury. The applicability of "injury per se" criteria. The "injured" that are easiest to prove are those which are set forth as specific quantifiable criteria in DOI Regulations. Examples are violations of water quality standards, levels in edible tissues in excess of FDA action levels, and levels in excess of those set by a State health agency in a directive to limit or ban consumption. However, recognize that contaminant levels in the surface microlayer cannot be directly compared to water quality standard/criterion levels. Also, note that exceeding a level set in a State fishing "advisory" is not the same as exceeding a limit contained in a consumption ban.

10. Feasibility of restoration. Recovered NRD monies can only be used for restoration. If restoration is not feasible, therefore, an NRD claim is not appropriate and resources should not be spent on assessment or quantification. Even where restoration is feasible, assessment is not appropriate where the cost will be "grossly disproportionate" to the value of the NRD claim.

Figure 8.6 Ten technical defense tips for potential natural resource damages.

agencies, but this may be possible especially if proactive restoration

- *Cleanup stage.* If a site is undergoing cleanup or remediation of contamination under CERCLA, an effort should be made to address cleanup and possible NRD problems simultaneously during all of the site investigation, planning, and implementation steps. By considering potential NRD problems during cleanup, it is possible to minimize the exposure to future damage claims and provide the basis for covenant not to sue.
- *Pursuit of a PRP claim stage.* At this stage, litigation and/or settlement alternatives loom. The company may wish to have a trustee show cause. Major tests must be demonstrated by a trustee to recover resource damages. For example, as discussed earlier, an NRDA claimant must demonstrate that:
    1. Releases are above thresholds required to inflict an injury
    2. There is a casual link between the release and injury
    3. Damages must be quantifiable and compensable

If any one of these criteria cannot be met beyond a reasonable doubt, then there is no basis for proceeding with all the other aspects of an NRDA claim. Further, there are a number of general defenses that have been successful in reducing claims by 90 to 95%, and this represents an option for dealing with any NRD claim. This approach, however, is not one that would necessarily enhance an already damaged public image of the company, and the latter should be strongly considered. The industries and the government federal facility complex are today paying a heavy price for yesterday's temporarily wise recalcitrance.

References

1. NRC, National Academy Marine Board Study, Washington, D.C., 1989.
2. Consulting Engineers Council of Metropolitan Washington, Guidelines, Environmental Site Assessments, October 5, 1989.
3. Tietenberg, T., *Environmental and National Resource Economics*, 3rd ed., 1999.
4. Gloyna, E.F. and Ledbetter, J.O., *Principles of Radiological Health*, New York: Marcel Dekker, 1969.
5. Thimann, K.V., *The Life of Bacteria*, 2nd ed., New York: Macmillan, 1963.
6. A.T. Kearney, Inc., Draft, Cleanup Levels for Restoration Engineering Study, 1991.
7. U.S. EPA, Risk Assessment Guidance for Superfund, Vol. 1, *Human Health Evaluation Manual (Part A)*, EPA/540/1-89/002, Washington, D.C., 1989.
8. U.S. EPA, Role of the Baseline Risk Assessment in Superfund Remedy Decisions, EPA OSWER Directive 9355.0-30, Washington, D.C., October 2002.
9. U.S. EPA, *Human Health Evaluation Manual*, Supplemental Guidance: Standard Default Exposure Factors, EPA OSWER Directive 9285.6-03, Washington, D.C., 1991.

chapter 9

Risk assessment and emergency response analysis

The current emphasis in environmental assessment and cleanup level development requires a thorough, comprehensive, and precise evaluation of risk, as well as an emergency response analysis that enables the site and the regulation to assess the risk inherent in accepting and implementing the cleanup. These evaluations are essential in determining the most appropriate remediation technique and the viability of cleanup standards and options.

Increasingly, EPA is also assessing the net health impacts resulting from remedial activity by subtracting the short-term risks due to remedial activity from the fatalities averted due to the reduction in long-term risks as a result of remedial activity. In this manner, the risk analyses evaluate the net health impacts at various cleanup levels and under various cleanup scenarios.

9.1   General discussion

EPA is increasingly weighing the merits of baseline risk assessments based on individual exposures vs. broader population-based risk analysis.

The overall goal of risk assessments is twofold:

1. Determine if a risk exists to the environment
2. Determine the level of risk to the environment, regardless of whether an individual exposure or population exposure technique is used

There are common elements to both techniques. A risk assessment may be used to:

1. Establish a baseline risk at a site

156   *Macroengineering: An environmental restoration management process*

There are two types of risk assessments:

1. A health assessment
2. An ecological assessment

As to the type and level of detail required for a risk assessment, it is dependent on site-specific conditions and objectives.

9.1.1   Health assessments

The objective of a health assessment is to provide:

1. A basis for determining levels of chemicals that can remain on site and still be protective of public health
2. An analysis of baseline risks that allows determination of the need for action with the site
3. A basis for comparing potential health impacts of various remedial alternatives
4. A consistent process for evaluating and documenting public health threats at sites

The health assessment components include:

1. Hazardous identification
2. Dose-response association
3. Exposure assessments
4. Risk characterization

Hazard identification, the first component of a health assessment, is the process of identifying which detected contaminants have inherent toxic effects and are likely to be of concern. It is based upon a review of site-specific monitoring and modeling information. The steps in hazard identification are as follows:

1. Determine the extent of contamination
2. Calculate statistical means
3. Evaluate nondetection and trace volume data
4. Determine the background and naturally occurring values
5. Select the contaminants of concern based upon concentration, toxicity, frequency of detection, sample location, and the preparation of the compound (chemical/physical)

A *dose-response assessment*, the second component of a health assessment,

Chapter 9: Risk assessment and emergency response analysis 157

1. Identification of exposure pathways
2. Estimation of exposure point concentrations for each selected pathway
3. Estimation of exposure dose for each selected pathway
4. Development of exposure scenarios

Exposure pathway is the key element in the exposure assessment and consists of four elements:

1. A source and mechanism of chemical release into the environment
2. An environmental transport medium (mechanism for the released contaminant to transfer from one medium to another)
3. A point of potential contact with humans and biota
4. A viable exposure route (air, groundwater, surface water, soil, food chain, etc.)

If all four elements are present, an exposure pathway is considered complete; if not, the potential risk is diminished significantly.

Exposure point concentration is defined as the amount of chemical in an environmental medium to which a person may be exposed. It can be expressed in either mass per unit volume [mg/l or mg (m^3)] or may be unit weight (mg/kg). Exposure point concentrations should be developed for each viable exposure pathway based on site sampling data or on modeling results.

The fourth component of a health assessment is *risk characteristics*. Chemical toxicity values, in conjunction with dose estimates for each of the various exposure pathways and population subgroups, can then be used to quantitatively estimate the carcinogenic health risks, as well as the noncarcinogenic health risks.

Risk assessment draws heavily upon the science of toxicology. *Toxicology* is the study of how toxic substances affect organisms. Central to these studies is the concept of *dose* and how it is expressed. If the difference between a toxic effect and no effect is the dose (and route of entry or exposure time), all chemical substances can produce harmful effects. Typical routes of entry are inhalation, ingestion, absorption, and injection. Dose can be recorded in units of mg/kg of body weight for the oral dose, cm^2 for the skin, dose, ppm, mg/m^3, and mg/L for the inhalation dose.

Toxicity is measured in terms of dose-response relationship. A toxicity test exhibits a dose-response relationship when there is a consistent mathematical and biologically plausible relationship between a proportion of individuals responding and a given dose for a given exposure period:

158   *Macroengineering: An environmental restoration management process*

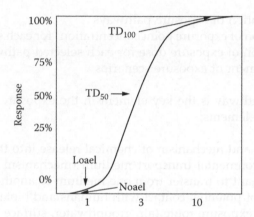

Figure 9.1 Dose-response curve (dose, arbitrary units, logarithmic scale). Routes of entry — inhalation, ingestion, absorption, and injection.

- NOAEL (no observed adverse effect level) — The concentration (or dose at which there is no adverse response in the population
- LOAEL (lowest observed adverse effect level) — The lowest concentration that causes an adverse response in the population
- TD_{50} (toxic dose 50) — The concentration (or dose) that produces the adverse effect in 50% of the population (LD_{50} for lethal dose, if killing the population)
- TD_{100} (toxic dose 100) — The concentration (or dose) that produces the adverse effect in 100% of the population (LD_{100} for lethal dose, if killing the population) as expressed in Figure 9.1, an example of a dose-response level

9.1.2   Ecological risk assessment

The objectives of an ecological risk assessment are to assess the probability of adverse biological and ecological effects related to site contamination. The assessment addresses the risk relative to past, present, and future site contamination impacts. The ecological risk assessment should play a key role in the development of cleanup criteria and the assessment of risk relative to remedial act alternatives.

There are five elements to an ecological risk assessment:

1. Site characterization and identification of potential receptors
2. Selection of chemicals, species, and endpoints
3. Exposure assessments

Chapter 9: Risk assessment and emergency response analysis   *159*

and species profiles. Relative to contamination, the researcher must establish the extent of contamination, the potential courses of contamination, and target the key contaminants of potential concern. In defining the habitat element to a given site, researchers need to characterize the species present, particularly endangered species and economically important species. It is also critical to establish what the potential indicator species are that reflect the overall ecological state of the site and provide a sound basis for the future site monitoring program.

The second ecological risk assessment element is the selection of chemicals, species, and endpoints. The basis for selection of chemical contaminant criteria is the persistence of the contaminant, its high bioaccumulation potential, the given chemicals toxicity, and the potential for elevated levels at a given site above naturally occurring background levels.

The basis for selection of species and endpoints criteria is:

1. Their respective importance to the ecological system
2. Their sensitivity
3. The availability of practical methods for measurement and predictiveness
4. The regulatory and trustee endpoint consideration

Potential general endpoints include organisms, populations, and community. For organisms, endpoints may be mortality rates, changes in growth, changes in behavior, changes in structural development, reproductivity, impairment, multigenicity, biochemical changes, and pathological abnormalities. From a population perspective, potential endpoints include species abundance, reproductive potential, and distribution. From a community perspective, potential endpoints for consideration include species composition, biomass, interspecies relationships, and extinction.

Exposure assessment is the third element of an ecological risk assessment. The objective of an exposure assessment is to identify:

1. The biological resources that are exposed to chemical contaminants
2. The significant pathways and routes of exposure
3. The magnitude duration and frequency of exposure

The components of the exposure assessment include:

1. Source characterization
2. Transport and fate analysis (i.e., migration mechanisms, spatial distribution, and temporal trends of contaminants)
3. Exposure scenarios
4. Uncertainty analysis

Toxicity assessment is the fourth element of an ecological risk assessment. The toxicity assessment objectives are:

1. Identify the potential toxic effects of the contaminants of concern
2. Identify the physical, chemical, and metabolic properties of each of the chemicals of concern
3. Determine the relationship between the amount of exposure to each chemical of concern and the resulting biological effect

The elements of the toxicity assessment are hazard identification, establishment of quantitative dose limits, and uncertainty analyses.

Hazard identification is based on the results of laboratory toxicity tests, field studies, and the quality review for the target endpoint in indicator species.

Quantitative dose limits is the response data and toxicological indices for given species and compounds. It is based on laboratory toxicity tests for individual chemicals and complex mixtures, as well as site-specific data.

The fifth and last element of an ecological risk assessment is risk characterization. Its objective is to determine the probability that adverse effects to the receptors of concern will result from the estimated exposure and to determine the degree of confidence in the risk estimate. The characterization of risk is based on:

1. The estimated risks for single chemicals, single species, and specific endpoints
2. The multiple chemical risk predictions
3. The distribution of estimated risks
4. The risks to communities and ecosystems
5. Uncertainty analysis

9.2   Standard Superfund baseline risk assessment practices

Computerized models are used to assess the potential doses and associated risks to the public from all significant exposure scenarios, media, and exposure pathways. The risk analysis focuses on estimating: (1) the change in risk to an individual under reasonable maximum exposure (RME) conditions and (2) the number of cancers per year in the exposed population. The methodology is consistent with that described in *EPA Risk Assessment Guid-*

Chapter 9: Risk assessment and emergency response analysis   161

The rural residential exposure scenario addresses long-term risks to individuals expected to have unrestricted general use of a site after cleanup. It assumes that individuals live on the site and are constantly exposed, both indoors and outdoors, to residual concentrations of hazardous constituents or radionuclides in soil through the maximum number of exposure pathways, including:

- External exposure to skin
- Inhalation of resuspended soil and dust containing contaminants
- Incidental ingestion of soil containing contaminants
- Ingestion of drinking water containing contaminants transported from soil to potable groundwater sources
- Ingestion of contaminated home-grown fruits and vegetables via take-up from the soil
- Ingestion of meat or milk containing contaminants taken up by cows grazing on contaminated plants and fodder
- Ingestion of locally caught fish containing contaminants

The commercial/industrial exposure scenario addresses long-term exposures and risks to commercial or industrial workers assuming a site was released with a restriction allowing commercial or industrial development. Under this scenario, the model assumes workers would be exposed to residual levels of contaminants in soil on the average.

- External exposure to the skin
- Inhalation of resuspended soil and dust containing contaminants
- Incidental ingestion of soil containing contaminants
- Ingestion of drinking water containing contaminants transported from soil to potable groundwater sources

The procedure for evaluating risks, whether from radionuclides or chemicals of concern, requires a number of distinct stages varying from initial site and source characterization, to assessing potential exposure routes and toxicity assessments.

Evaluation of risks must address the impact from a variety of sources. The doses may result from direct exposure or emissions in air and/or water. The exposure may be of a relatively short- or long-term duration and may be the result of operational activities or major accidents. The approach for evaluating the dose impact should include the following steps:

162   *Macroengineering: An environmental restoration management process*

- Human health assessments for contaminant concentration in environmental media
- Environmental fate and transport
- Pathway analysis and quantification of uptake
- Toxicology and radiological health physics
- Statistical uncertainty analysis
- Identification of contaminants and pathways of concern

Relative to human health assessments for contaminant concentrations in environmental media, there are three steps: (1) exposure scenario development, (2) determination of critical pathways, and (3) identification of critical population groups.

Environmental fate and transport support activities involve assessing the critical transport media for each contaminant of concern. This involves: (1) determination of migration potential, (2) determination of release rates, (3) assessment of appropriate modeling techniques, and (4) interpretation of source concentrations.

Pathway analysis and quantification of uptake involves: (1) quantification of exposure potential, (2) identification of critical population groups, (3) determination of exposure concentrations based on activity/lifestyle, and (4) determination and prioritization of exposure mechanisms.

Toxicological and radiological health physics studies involve: (1) evaluation of dose and risk to selected exposure groups, based on concentrations in environmental media, (2) generating dose and risk evaluation based on intake and exposure routes, and (3) identification of hazard potential of radionuclides of concern.

The elements of statistical uncertainty analysis includes: (1) evaluation of statistical uncertainties using statistical sampling techniques such as Monte Carlo and Latin Hypercube, (2) development of strategies for performing uncertainty and sensitivity analysis on the models and processes involved in characterizing risk, and (3) generating a statistical evaluation and probabilistic determination of risk.

The approach for completing an identification of contaminants and pathways of concern is to generate an identification matrix to determine the relative hazards of contaminants with respect to: (1) environmental media, (2) exposed population groups, and (3) risk vs. time profiles.

Risk assessment activities should be done in conjunction with regulatory strategy preparation and should be clearly reflected and integrated into the detail design process. Detail risk assessment studies also provide both guidance to system design and also a firm basis for establishing the cost/benefit

Chapter 9:   Risk assessment and emergency response analysis   *163*

- Hazardous substances present or releases to relevant media (e.g., air, groundwater, surface water, and soil)
- Exposure pathways and contribution of each pathway to the overall risk
- Populations at risk
- Intrinsic toxic properties of the hazardous substances
- Extent of present exposure and expected exposure from site operations
- Determination of risk zones and the probability of harm occurring within these zones to workers, the public, and the environment

The risk assessment process relies upon an evaluation and interpretation of the present data concerning releases at the site. The data to be collected and reviewed includes demographic, geographic, physical, chemical, and radiologic (if appropriate) data that has been generated by monitoring programs over a period of years.

Accepted EPA models should be used to estimate the impact of any releases to air, water, and soil quality in the site area. The models should use statistical summaries of meteorological data, river data of other available information on the site area in estimating the rate of transport, transformation, and deposition of radioactive and chemical pollutants in the various media systems. The models that would be used include:

- Industrial source complex long-term model (long-term air quality impact)
- Single source Gaussian dispersion algorithm (air quality from a single industrial source)
- Seasonal soil compartment model (movement of materials in soil)
- Analytical transient-dim model (movement of materials in groundwater)
- Exposure analysis model (simulates the fate of materials in a surface water ecosystem)

The output of these models would then be compared to that of models that may have been utilized at various sites. If other models are considered more appropriate, then they would be utilized.

9.3   Population risk analysis

As described in the earlier section, currently EPA's risk assessment guidance for Superfund (RAGS) assesses risk to a reasonable maximum exposed (RME) individual. However, consideration is being given to consider using population risk as a risk descriptor in describing and communicating risks. Specifically, "The Policy for Risk Characterization" issued by Administrator Carol Browner in March 1995 states that:

In addition, President Clinton's Executive Order 2866 on regulatory planning and review included provisions designed to promote the use of risk analysis in making regulatory decisions in which benefits justify costs.

The macroengineering approach favors an aggressive use of population risk to define site restoration programs, recognizing that the latter may result in the selection of more cost-effective (and implementable) remedies for site remediation. Furthermore, a population risk approach may generate more statistically sound data.

The standard baseline risk assessments discussed in Section 9.2 summarize the relative contribution that each substance makes toward the total RME cancer risk and hazard index (noncancer health risk) for each exposure medium. When calculating the RME for individual risks, EPA uses the 95% upper confidence limit on the arithmetic mean concentration (95-UCL) as the assumed source of exposure for each substance. The concentration is used to provide an upper bound to individual exposures and risk. However, when estimating population impacts, it is more appropriate to use a measure of central tendency such as an arithmetic average or geometric mean concentration because most individuals in the population are not experiencing upper limit exposures over their lifetime. Whereas the total contaminant intake for a population is the sum of all individual intakes, some of the individual intakes will be somewhere in between, depending upon the frequency distribution of contaminant concentrations. Almost all frequency distributions of contaminant concentrations are log-normal and in this type of distribution, the geometric mean best represents the central tendency and is the most frequent value for each environmental medium. The difference between the geometric mean and 95-UCL may be quite significant.

Assessment of population risk has been an integral part of the radiation protection field for over 20 yr. Population risk is incorporated into the regulatory framework of all federal and state radiation protection agencies and is based on the recommendation of the National Council on Radiation Protection and Measurement (NCRP) and the International Commission on Radiological Protection (ICRP).

Many of the technical issues and methodologies used in the radiation protection field are common to Superfund. Population risk assessments for hazardous waste sites can be performed for health (human) and ecological assessments, using relatively simple techniques and readily available information. In fact, most of the relevant information and data can typically be found or developed from a site's RI/FS and baseline assessment. For example, population risk model algorithms can be fashioned from RAGs human health evaluation equations. The models assure that population risks are

Soil exposure models should be developed under the assumption that contamination is exponentially depleted from surface soils through degradation or leaching processes. Contaminant depletion is usually not accounted for in most risk assessments, but many organic compounds can degrade or be depleted from surface soils by biological and chemical processes. In addition, elements can be removed from surface soil layers through erosion and leaching. Over a lifetime, this depletion can be significant, resulting in exposures that are one to two orders of magnitude less than when no loss is assumed. Both soil ingestion and soil inhalation should be addressed.

Individuals living or working on the property can ingest soil inadvertently. Model parameter values can be developed that are both specific hazardous-substance dependent (soil concentration, loss rate constant, slope factor, etc.) and independent (soil ingestion rate, residual time, lifetime exposure, body weight, population, etc.). Except for the soil loss rate constant and the total number of individuals in the population, the latter values can be taken directly from the baseline risk assessment. Loss rate constants will vary for organics, chemicals, and metals and can be obtained from the scientific literature. The number of individuals in the exposed (industrial or residential) population can be estimated from census reports and can also be found in RIs.

The soil exposure model should also address the potential of individuals living or working on the property, inhaling contaminated-surface soil particles suspended by wind action or vehicular traffic. Two model parameters need to be developed to assess this exposure pathway — soil dust loading in air (DL) and inhalation rate. The DL value can be empirically derived, accounting for all factors that influence soil suspension, such as particle size and wind velocity. The inhalation rates can be typically taken from the baseline risk assessments.

Individuals living on or near the property can also ingest contamination from garden crops grown in contaminated soil. To calculate the total number of concerns in the population, two additional parameters should be developed:

1. The concentration ratio (CR), which is substance dependent
2. The daily crop intake (CI) for crop produce. The CI value can be taken from the soil screening guidance technical background document (EPA, 1996)

The basic approach focuses on the specific number of individuals residing on or near the contaminated land. However, there may be circumstances

166   *Macroengineering: An environmental restoration management process*

and should be evaluated under a thorough population risk analysis. It is advisable to assume no loss of contamination over time if the site soil represents a continuous source of contamination to the groundwater. Parameter values can typically be taken directly or derived from data in the RI/FS report. The estimate of the number of exposed workers or residents will be dependent on establishing the extent of the groundwater plume and the industrial and residential densities. However, there may be a need to address the more challenging scenario of groundwater being a resource to the municipal water supply.

In summary, population risks can be used to determine the most effective standard or remedy based on health benefits and other factors, such as social, economic, and technical feasibility. It can also be a crosscheck on the reasonableness of the current Superfund approach. As discussed earlier, the Superfund program presently estimates individual risks under their baseline risk assessment approach (i.e., RME) and uses this quantity along with other criteria to determine the need for remedial action and the extent of remedial action required. Typically, individual cancer risks greater than 10^{-4} require some form of site action. The Superfund program, however, does not use a qualitative assessment of population risk to help identify the most effective remedy. The overall risk reduction is usually ambiguous among the different remedies being considered.

Quantifying the overall risk reduction of a particular remedy in terms of population risks can also promote a balancing between risks from site contamination and the risks from implementing the remedial action (i.e., cancer mortalities from site contamination vs. mortalities expected from construction, worker exposure, and traffic accidents during the cleanup).

Whereas the individual cancer risk for residential exposure is almost always greater than individual worker risk (because of factors such as intake rates, time on the property, and the large number of exposures), the reverse is typically found under a population risk analysis. This is due to the fact that the industrial population, assumed to be working on the property, is typically much greater than the number of residents living on or near the site. The greater the number of individual workers compared to the number of affected residents more than complements for the greater individual residential time frame of exposure.

Thus, the results of the population risk assessment can provide additional insight for the remedy selection process that cannot be realized by analyzing individual risk alone.

9.4   *Emergency response analysis*

Chapter 9:   Risk assessment and emergency response analysis   **167**

associated with the macroengineering approach environmental restoration; it is not intended to be used as the definitive analysis for any given site.

At this time, there is no indication that the macroengineering approach presents any greater hazard to the community than the microengineering approach. However, due to the potential lack of documentation on past waste handling practices and the emphasis on real-time characterization at the face, and emergencies involving unknown waste types in buried containers must be considered when evaluating risks at the potential site. The scenarios discussed in the following are applicable to the macroengineering approach or any other approach in which drums or containers may be punctured during sampling or removal activities.

It is probable that all the hazardous waste substances at a given site could produce an adverse effect that manifests itself as an alteration of a normal biological function. Risk is the probability that one of the substances will produce harm under certain conditions. Safety, in the emergency response analysis context, is the reciprocal of risk, or the probability that substances will not produce harm under the same conditions. Thus, when ultimately determining the risk or safety of the present approach, the critical factor will not necessarily be the intrinsic hazard potential of the substances *per se*, but the safety precautions that have been taken to decrease the potential harm from that substance.

It is assumed that the greatest hazard potential to the nearest community will be via air in the form of a vapor, gas, or dust cloud or via an explosion of a munitions container. Using these assumptions, worst-case scenarios for representative compounds should be conducted to indicate the size of the hazard zones that would have to be considered in emergency planning for the present approach.

The ARCHIE model series was developed by the Environmental Protection Agency (EPA), the Department of Transportation (DOT), and the Federal Emergency Management Administration (FEMA) for use in hazard analysis. ARCHIE is the recommended model in several states as part of the emergency response analysis requirement in permit requests. It should be stressed that the assumptions are considered to be worst case, i.e., the situation is assumed to be the worst that can occur based upon the information. As such, severe combinations of factors (i.e., temperature, wind, and others) have been fed into the model than would normally be encountered at a given site that results in exaggeration of the size of hazard zones.

In the example presented, the condensed phase explosion model assesses the physically destructive impact of a potential explosion. Table 9.1 is an

Table 9.1 Condensed Phase Explosion Effects

Distance from explosion (ft)	Expected damage
10,676	Occasional breakage of large windows under stress.
1,505	Some damage to ceilings; 10% window breakage.
562–974	Windows usually shattered; some frame damage.
144–562	Range serious/slight injuries from flying glass/objects.
339	Partial collapse of walls/roofs.
259–339	Nonreinforced concrete/cinder block walls shattered.
116–3,090	Range 90–1% eardrum rupture among exposed population.
292	50% destruction of brickwork.
216–259	Frameless steel panel buildings ruined.
189	Wooden utility poles snapped.
156–189	Nearly complete destruction of wooden structures.
128	Probable total building destruction.
77–106	Range for 99–1% fatalities among exposed populations due to direct blast effects.

Note: The explosion is assumed to take place on or near the ground.

TNT can be placed in a 55-gal drum. Table 9.1 represents the various hazard zones if such an amount of TNT were detonated.

The case model assumed that the location of the explosions is at ground level and that the surrounding area is flat and without obstacles. In cases in which drums would be encountered below the ground, the explosion would produce smaller hazard zones than those predicted. In the example, this model indicates no emergency would exist past 1000 ft from the explosion and, therefore, would be no threat to the community.

The purpose of the vapor dispersion model is to provide an estimate of the dimensions of the initial downwind zone that may require protective action in the event of a hazardous material discharge due to the accidental discharge, emission, or release of a toxic gas or vapor to the atmosphere.

The worst case for vapor dispersion would be the substance with the highest vapor pressure and lowest IDLH (immediately dangerous to life and health) dose. This is because the model assumes that all the contents of the drum will be released to the atmosphere in 1 min, and vapor pressure, thus, does not play a significant role. This assumption creates an extremely conservative model because, even with the drum and ambient temperatures at 110°F, the entire contents of the drum would not vaporize in 1 min unless the rupture

Chapter 9: Risk assessment and emergency response analysis 169

possibility of tritium being present at the waste sites. As far as the mechanism of movement in the example through the natural environment is concerned, the activity level of a radionuclide is irrelevant, except in a few cases in which recoil processes contribute to release mechanisms. Because radioisotopes have chemical properties identical to those of their stable homologs, their movement will essentially follow that of the stable element. Because hydrogen is nontoxic, except as an asphyxiate, the model shows a zone in which hydrogen would achieve a concentration of 1000 ppm. This model assumed that the temperature inside the pipe as well as the external temperature was 110°F and that the entire content of the pipeline was released in 1 min. It should be noted that if tritium is substituted, the zones would be smaller because tritium has a higher molecular weight than hydrogen (see Table 9.2).

The results in Table 9.2 indicate that in the unlikely event that this volume of gas under 100 psia was released in 1 min, the 1000-ppm concentration zone would be less than 1000 ft and would not pose a threat to the area. Again the zone would probably be less, because any release would be below ground level and not at 0 ft as assumed in the model.

Inhalation exposure due to contaminated airborne dust may occur. Inhalation exposure to contaminants depends on the quantity of dust in the air and its particle size. In the example, potential exposure to the compounds via dust is mitigatable because of the wetting and covering procedures

Table 9.2 Example of Hydrogen Gas Dispersion Analysis Results

| Downwind distance | | Ground-level concentration | Initial zone |
ft	mi	(ppm)	width (ft)
100	0.02	52,292	73
153	0.03	22,861	120
205	0.04	12,948	150
257	0.05	8,411	190
309	0.06	5,947	230
361	0.07	4,454	270
413	0.08	3,476	310
465	0.09	2,800	340
517	0.1	2,311	380
569	0.11	1,945	420
621	0,12	1,664	460
673	0.13	1,443	490

170   *Macroengineering: An environmental restoration management process*

Table 9.3 Dust Dispersion Analysis Results

Downwind distance		Ground-level concentration	Initial zone
ft	mi	(ppm)	width (ft)
100	0.02	565	73
194	0.04	155	150
288	0.06	73	210
382	0.08	43.3	280
476	0.09	29	350
569	0.11	21	420
663	0.13	16	490
757	0.15	12.7	560
851	0.17	10.41	620
945	0.18	8.7	690
1038	0.20	7.4	760
1132	0.22	6.4	830
1226	0.24	5.6	900
1320	0.25	4.9	970
1414	0.27	4.4	1

Note: Usually safe for <1-h release. Longer releases or sudden wind shifts may require a larger width or different direction for the evacuation zone. Source height specified by the user for this scenario was 0 ft.

planned during the macroengineering approach. In the example, the dust dispersion model assumes that 10 lb/min of dust is generated, and that 20% of it will be small enough to be respired by a human, and that 50% will be totally absorbed by the person (see Table 9.3).

In the example, the model assumes a 10-mg/m^3 nuisance dust level to establish the zones. These results indicate that at dust levels used in the model, the hazard zone is within 1500 ft and would not include the community.

Bibliography

1. Choices in Risk Assessment: The Role of Science Policy in the Environmental Risk Management Process, Prepared for Sandia National Laboratories, for the Dept. of Energy (DOE), 1994.
2. U.S. EPA, Office of Acid Deposition, Environmental Monitoring, and Quality Assurance. The Total Exposure Assessment Methodology (TEAM) Study: Summary and Analysis: Vol. 1, EPA/600/6-87/002a, Washington, D.C., 1987.

Chapter 9:   Risk assessment and emergency response analysis   *171*

5. DOE, *A Manual for Implementing Residual Radioactive Material Guidelines*, DOE/ CH8901, Washington, D.C., June 1989.
6. U.S. EPA, Office of Air and Radiation, Technical Support Document for the Development of Radionuclide Cleanup Levels for Soil, Review Draft, Washington, D.C., September 1994.
7. U.S. EPA, Office of Solid Waste and Emergency Response, Soil Screening Guidance: Technical Background Document, EPA/540/R-95/128, Washington, D.C., May 1996.
8. U.S. EPA, Office of Research and Development, Update to Exposure Factors Handbook, Draft Report, Washington, D.C., August 1996.
9. U.S. EPA, Office of Emergency and Remedial Response, Population Risk Analysis for Superfund Sites Using Simple Techniques and Readily Available Information, prepared by SC&A, March 1998.
10. Hoffman, F.O. and Bates, C.F., III, A Statistical Analysis of Selected Parameters for Predicting Food Chain Transport and Internal Dose of Radionuclides, NUREG/CR-1004, Washington, D.C., 1979.
11. Schaffer, S.A., Environmental transfer and loss parameters for four selected priority pollutants, Proceedings of the National Conference on Hazardous Waste and Environmental Emergencies, Cincinnati, OH, May 1985, pp. 145–149.

chapter 10

Establishing project hazards and safety control measures

There are numerous radiological, chemical, physical, and environmental hazards potentially present at Sites. These hazards, if not properly controlled, can cause harm to project personnel, visitors, and the public. The anticipated hazards at the project and the recommended control measures need to be identified and addressed in detail in site health and safety plans per OSHA regulations for hazardous work operations and emergency response (HAZWOPPER 29 CFR 1920.120).

Historical information and site characterization data can be used to indicate the presence of contaminants and other hazards of concern. Typically, there is potential for exposure to personnel through various routes (dermal contact, inhalation, ingestion, and injection). Controls must be specified in site health and safety plans to reduce the risk of these potential exposures.

A brief definition of important inhalation exposure terms is provided in the following:

- *Threshold limit value — time-weighted average (TLV-TWA)*. Airborne concentrations of substances, generally expressed as an 8-h TWA and represent conditions under which it is believed that nearly all workers may be repeatedly exposed day after day for a 40-h work week without adverse health effects. TLVs are guidelines for occupational exposures established by the American Conference of Governmental Industrial Hygienists (ACGIH, 1998), and should be used only on controlled sites in which contaminants and concentrations are well known.
- *Threshold limit value — short-term exposure limit (TLV-STEL)*. The concentration to which it is believed that workers can be exposed

174   *Macroengineering: An environmental restoration management process*

impair self-rescue, or materially reduce work efficiency, provided that the daily TLV-TWA is not exceeded. A STEL is defined as a 15-min TWA exposure that should not be exceeded at any time during the work day even if the 8-h TWA is within the TLV-TWA. Exposures above the TLV-TWA up to STEL should not be longer than 15 min and should not occur more than four times per day. There should be at least 60 min between successive exposures in this range.

- *Recommended exposure limit* — The up-to-10-h/d TWA exposure limits recommended by the National Institute of Occupational Safety and Health (NIOSH).
- *Immediately dangerous to life or health (IDLH)* — Concentration that poses an immediate threat to life or produces irreversible, immediate debilitating effects on health (American National Standards Institute). NIOSH defines IDLH as air concentrations that represent the maximum concentration from which, in the event of respirator failure, one could escape within 30 min without a respirator and without experiencing any escape-impairing or irreversible health effects.
- *Permissible exposure limit (PEL)* — The 8-h TWA, STEL, or ceiling concentration above which workers cannot be exposed. Enforceable standards by OSHA.

10.1   *Inorganic chemicals*

Various inorganic chemicals, specifically metals, can be considered toxic, and some are identified as being carcinogenic. Detect analysis for each contaminant of concern should be presented in the health and safety plan. For example, arsenic is a toxic, gray, brittle metal that may injure multiple organs. Acute injury usually involves the blood, brain, heart, kidneys, and gastrointestinal tract. The bone marrow, skin, and peripheral nervous system may develop chronic toxicity after acute or chronic exposure. Thus, an acute ingestion may cause both acute and chronic syndromes. The ACGIH has listed arsenic as an A1, confirmed human carcinogen. (PEL: 0.010 mg/m^3, IDLH: 5 mg/m^3, TLV-TWA 0.010 mg/m^3) TLV basis-critical effects: cancer (lung, skin).

10.2   *Organic compounds*

Organic compounds (hydrocarbons) may also be present as contaminants in the soil. Additional information about these chemicals should be found in

The greatest carcinogenic effect is at the point of contact (i.e., lungs, skin, and stomach). Skin disorders may also result due to high-concentration exposures. Exposure limits have not been established for many specific PAHs in this large group of compounds.

10.3 Operational chemicals/hazard communication program

Hazardous chemicals will be brought to the project site for use in activities supporting the planned work. These chemicals are used as fuels in operating heavy equipment, glues for welding pipes, painting, etc. The use of operational chemicals is regulated by OSHA under the *Hazard Communication Standard* (29 CFR 1910.1200). Air monitoring must be performed as needed to assess exposures resulting from their use. MSDSs for operational chemicals must be kept on file in the project office trailer, and in inventory list of the anticipated operational/laboratory chemicals (*Hazardous Chemical Inventory List*) for use at the site must be maintained at the Health and Safety (H&S) field office.

Other important terms and concepts of chemical hazards include fire/flammability and flammable or explosive limits. For fire/flammability to be a concern, three elements that must be present are fuel, heat, and oxygen.

Flammable or explosive limits are measured in terms of a flammable range bounded by the lower explosive level (LEL) or lower flammable limit (LFL) and the upper explosive level (UEL) or upper flammable limit (UFL). Figure 10.1 provides an example.

10.4 Radiological hazards

Radioactive materials present unique health and safety concerns and should be recognized as such through the presence and input of a radiation health and safety operation officer.

Alpha particles are normally not considered an external dose hazard for workers, as an alpha particle is stopped by the dead layer of the skin. However, once the alpha particle is inside the body (typically, the lung would be the initial entry), the particle's energy is deposited in the living tissue.

Term

100%

Too rich

176   *Macroengineering: An environmental restoration management process*

This deposited energy may result in several things happening. The cell may die, continue to function normally, change function, or become cancerous. Therefore, it is prudent to take reasonable precautions against entry of alpha-emitting radioactive materials into the body.

Other radioactive emissions that are given off during the radioactive decay of the progeny radionuclides are beta particles and gamma rays. Beta particles are the equivalent of an electron, except they originate in the nucleus of the atom. Energy of beta particles varies widely, with initial beta maximum energies ranging from tens of thousands of electron volts (eV) to 2.3 MeV. On an atomic scale, beta particles are small so they can travel much farther in air and matter than alpha particles. A rule of thumb for beta particles is that a 1-MeV beta will travel about 11 ft in air.

Beta particles are considered an external and internal radiation hazard. The energy of the beta particles can be deposited externally in the skin or internally if the radioactive material gets inside the body. When large amounts of beta particles interact with the skin, they can cause reddening of the skin, much similar to sunburn. Internally, the beta particle energy will be deposited in living tissue. However, there is less energy deposited per cell than with an alpha particle, so there is less risk that the energy may result in changes to the cell. As with alpha particle energy deposition, the fate of the cells remains the same for energy deposited by a beta particle.

Gamma rays are high-energy, short-wavelength rays. With the exception of the higher energy, they are similar to light rays. These high-energy rays can travel long distances in air and in matter. Unlike alpha and beta particles, they have no well-defined range in matter. They can travel through material without depositing any energy or they may be completely absorbed. However, as they can travel large distances in air and matter, there is little energy deposited per unit path length or in any one cell. For this reason, gamma rays are considered to cause whole-body irradiation and are not considered an internal hazard, because gamma rays emitted inside the body may not deposit any energy as it travels through the body.

Neutron particles are neutral particles emitted from the nucleus. Neutrons have one fourth the mass of an alpha particle. Neutron decay in beta particles can produce alpha, beta, and gamma particles when it impacts hydrogen or nitrogen atoms. All four forms damage living organisms by imparting energy that ionizes water molecules in cells, allowing for the formation of hydrogen peroxide inside the cell. Hence, these kinds of radiation are referred to as *ionizing*. Ionization can disturb normal cellular function, resulting in sickness or can cause death.

Given this information, the radioactive contaminants represent an exter-

The beta and alpha particles also represent an internal dose concern when these materials enter the body through inhalation, ingestion, or injection. Once in the body, there are few methods available for removal, so the energy of the particles is deposited in the internal tissue, thus giving dose to the organ in which the material is deposited.

The rules that govern worker exposure to radioactive materials are found in 10 CFR 19 and 20. This will include development of an as low as reasonably achievable (ALARA) program and a radiation work permit (RWP) that briefly describes the scope of the work to be performed, the radiological conditions within the work area, and lists proper protective clothing requirements and monitoring requirements.

10.5 Unexploded ordnance

Unexploded ordnance (UXO) represents a unique health and safety issue and must be addressed through strict site control and personnel requirements. Areas in which UXO is suspected should be screened by certified explosive ordnance (EOD) personnel and who provide ongoing safe handling direction. Central to this effort is the development of an explosive safety submission (ESS) plan.

The ESS outlines the safety aspects for site characterization, remedial design, and remedial action at sites suspected to have UXO. All activities involving work in areas potentially contaminated with UXO must be conducted in full compliance with U.S. Army Corps of Engineers (USACE), Department of the Army (DA), and Department of Defense (DOD) requirements regarding personnel, equipment, and procedures.

The ESS identifies areas of concern at the site and the procedure to be used to mitigate and address UXO impacts. This includes development of quantity–distance (Q/D) maps that establish the removal depth and test depth as well as the Q/D for the storage of demolition explosives.

The ESS should also identify the expected amounts and types of ordnance and explosive (OE). The most probable munition (MPM) for each area of concern will be the OE item causing the worst-case scenario. However, if an OE item with a greater fragmentation distance is found, the Q/D arcs must be adjusted.

The methods of handling and disposing off OE should be specified in the ESS. This includes establishing Q/D, the public withdrawal distance (PWD) for all unrelated personnel for an unintentional detonation, and the personnel separation distance (PSD) for all personnel (related or unrelated) for intentional detonations. These exclusion zones (EZs) must be established around the grid being cleared of OE items. It should be understood

the magazines should have a lightening protective system in accordance with DOD 6055.9-STD and NFPA 780.

The ESS should also provide the safety arrangements for ensuring crew safety from lightning (i.e., shutdown if lightening is within 10 mi of crew operation, monitoring of weather channel reports, etc.).

The type, amount, class, and net explosive weight (NEW) stored in the magazines should be listed in the ESS along with required safe separation distance, based on DOD ammunition and explosives safety standards (DOD 6055.9-STD,E.2.c).

Planned or established demolition areas should be identified and described as well as "footprint" areas. There are three types of footprint areas: blow-in-place (BIP), OE collection points within a search grid, and consolidated shots within a search grid.

The BIP method is used for OE items not safe to move. The demolition locations will be confined to the boundaries of each subarea. Demolition sites will exist where UXOs are found and detonated. The location of UXO, which must be detonated in place, cannot be predicted, and they could occur at any point on the site. All UXO that are detonated in place must be well documented and the position indicated on the site map. With the ESS, tables must be developed to deal with intentional detonation and identify the withdrawal distances for all personnel for munitions and/or explosives expected to be encountered during UXO operations. If an OE not listed in the site ESS safe separation distance is encountered, its withdrawal distance requirements shall be determined in accordance with determination of appropriate safety distances on OE project sites, OE center of expertise (CX) interim guidance document 98-08. Until distances are determined by determination of appropriate safety distances on OE project sites, OE CX interim guidance document 98-08, the default distances in DOD 6055-9-STD (Chapter 5, Paragraph E.4.a) shall be used.

In-grid consolidation shots are areas within the search grids where OE items that are safe to move are collected and destroyed through explosive detonation. The size of the EZ and the set up of the shot will be accomplished in accordance with the DDESB-approved method (i.e., procedures for demolition of multiple rounds).

If a suspect chemical warfare materials (CWM) is encountered or a military ordnance item cannot be positively identified, the site safety officer should stop all operations on-site, evacuate personnel, secure the site, and notify the designated chemical warfare safety specialist of the Army Corp of Engineers. Once the specialist has confirmed the item to be CWM, the specialist and site safety officer must request the support of the newly designated U.S. Army Technical Escort Unit.

- Noise
- Slips, trips, and falls
- Fires, explosions, and hot work
- Use of ladders and scaffolding
- Use of small tools
- Use of heavy and mechanized equipment
- Operation of motor vehicles
- Materials handling
- Hazardous energies (electrical, mechanical, and pressure)
- Excavation
- Demolition
- Dust
- Railroads

10.7 Personal protective equipment

When engineering and administrative controls are not feasible or not adequate to protect personnel from the hazards associated with project activities, personnel practice equipment must be used.

10.7.1 Respiratory protection

When deemed necessary, a respiratory protection program should be implemented that is compliant to the requirements of 10 CFR 20 Subpart H, *Respiratory Protection and Controls to Restrict Internal Exposure in Restricted Areas*, and EM825-1-1 06.E.07, *Respiratory Protection and Other Controls*. Respiratory protection equipment must be NIOSH-approved, and respirator use must conform to ANSI Z88.2 and OSHA 29 CFR 1926.103 requirements. That details the selection, use, inspection, cleaning, maintenance, storage, and fit testing of respiratory protection equipment.

10.7.2 Levels of protection

PPE is used as a last line of defense to control employee exposure to hazardous chemicals. PPE must be selected based on the hazards identified, must be appropriate for the degree of hazard, and employees must be trained on the selection, use, care, and advantages/disadvantages of the PPE.

Eye protection: In areas where there is the potential for flying objects,

180   *Macroengineering: An environmental restoration management process*

Hand protection: Anywhere there is the potential for cuts, abrasions, punctures, chemical burns, thermal burns, or harmful temperatures, hand protection must be offered.

- Fit
- Types of gloves
- Barrier creams

Chemical protective clothing: Required when the employee has potential exposure to airborne contaminants, splashing, spilling, or other activities in which full-body contact is possible, chemical protective clothing must be worn.

- Aprons/bibs
- Suits
- Levels of protection (see Figure 10.2) — level A, level B, level C, and level D

Respiratory protection: Employees with potential exposure to dust, fumes, mist, vapors, or sprays must be provided respiratory protection if engineering controls or administrative controls are not feasible.

- Dust masks
- Air-purifying respirators
- Supplied air

Hearing protection: Employees exposed to continuous noise at or above 85 dBA for an 8-h time-weighted average (TWA) must be provided with hearing protection and enrolled in a hearing conservation program.

- Earplugs
- Earmuffs
- Attenuators

10.7.2.1   Level A protection

Level A protective equipment, if utilized, shall consist of an enclosed self-supplied air respirator with personnel in a chemically compatible enclosed (i.e., moon suit) working suit and boots with an airtight splash shield assembly (see Figure 10.2). Level A should always be used when the expected concentrations are at or near IDLH.

Chapter 10: Establishing project hazards and safety control measures   *181*

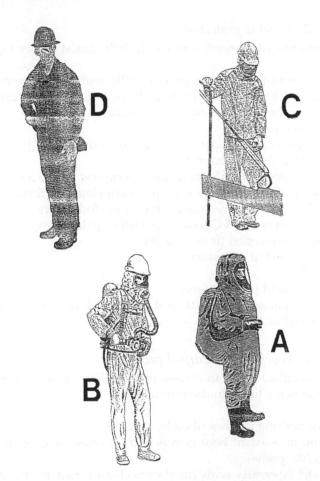

Figure 10.2 Levels of protection — level A, level B, level C and level D.

- Steel-toed boots
- Chemical-resistant boot covers and/or outer boots (as selected by a CIH)
- Tyvek® coveralls with hoods or an equivalent protective garment (as determined by the SSHO and RSO), elastic wrists and ankles (or equivalent cloth/synthetic fiber)
- Acid gear, splash suit, rain gear, etc. (as determined by a CIH)
- Nitrile, latex, or vinyl gloves (inner) and/or cloth liners
- Outer gloves (as selected by a CIH)

182   *Macroengineering: An environmental restoration management process*

10.7.2.3 Level C protection

Level C protective equipment, if utilized, shall consist of (see Figure 10.2):

- Full-face air-purifying respirator (APR) with NIOSH-approved combination high-efficiency particulate air/organic vapor cartridges
- Work clothing as prescribed by weather
- Steel-toed boots
- Chemical-resistant boot covers and/or outer boots (polyvinyl chloride (PVC)/latex/neoprene)
- Tyvek® coveralls with hoods (as determined by the SSHO and RSO), elastic wrists and ankles (or equivalent cloth/synthetic fiber)
- Nitrile, latex, or vinyl gloves (inner) or cloth liners
- Nitrile gloves or PVC (outer) or leather palm gloves
- Hearing protection (if necessary)
- Cooling vest (if necessary)
- Hard hat
- Splash shield (if necessary)
- Openings at ankles, wrists, and hoods shall be taped (as directed by the SSHO or RSO)

10.7.2.4 Level D — modified protection

Level D — Modified PPE can consist of the minimum level D plus any of the additional items listed in the following:

- Work clothing as prescribed by weather
- Chemical-resistant boot covers and/or totes (or equivalent) (PVC/latex/neoprene)
- Tyvek® coveralls with hoods (as determined by the SSHO and RSO), elastic wrists and ankles (or equivalent cloth/synthetic fiber)
- Nitrile or vinyl gloves (inner) or cloth liners
- Nitrile or PVC gloves (outer) or leather palm gloves
- Hearing protection (if necessary)
- Splash shield (if necessary)
- Cooling vest (if necessary)
- Openings at ankles, wrists, and hoods shall be taped (as directed by the SSHO or RSO)

10.7.2.5 Level D protection

Level D protection is the minimum level of protection that will be used at

Chapter 10: Establishing project hazards and safety control measures   **183**

- Hard hat
- Splash shield (if necessary)
- Leather work gloves (as necessary)

10.7.3 Monitoring and medical surveillance

Monitoring is done to verify the absence or presence of hazardous materials in the work environment. A medical surveillance is performed to verify the absence or presence of employee exposure to hazardous chemicals.

Monitoring: Monitoring can be done both for area contaminants and for employee exposure (personal monitoring).

- Area monitoring: looking at atmospheric conditions (explosive levels, oxygen levels, organic vapors, etc.)
- Personal monitoring: looking for potential exposure to employees
- Background monitoring
- Periodic monitoring
- Postincident exposure monitoring

Measurement instruments: There are two general approaches used to identify and/or quantify airborne contaminants:

1. On-site use of direct-read instruments
2. Lab analysis of samples taken

The advantage of direct-read instruments is that it provides real-time data.

Disadvantages of direct-read instruments include their limits in detecting/measuring of specific classes of chemicals. They are not typically designed to detect <1 ppm and are subject to interference problems. Direct-read instruments are:

- Combustible gas meter
- Oxygen meter
- Flame ionization detector
- Photoionization detector
- Colorimetric tubes
- Gas-specific instruments
- Radioactivity detectors
- Particulate detectors

Other various monitoring devices typically used include:

184   *Macroengineering: An environmental restoration management process*

- Geiger–Mueller pancake probes
- NaI scintillation meter

10.8 Site control and work zones

Site control requires the designation of work zones at the project as required by 10 CFR 20 Subpart J, specifically 1901 *Caution Signs*, 1902 *Posting Requirements*, and 1904 *Labeling of Containers*. These requirements are mirrored in EM 835-1-1 06.E.08, *Signs, Labels, and Posting Requirements*.

Entrance to posted contamination areas or radiation areas will be through designated entryways. For personnel, these entry areas will be noted as a contamination reduction zone (CRZ), where PPE will be doffed as the worker leaves the contaminated area. For large equipment, an equivalent CRZ will be designated.

For radiation contamination, a nuclear regulator commission (NRC) form 3, "notice to employees," shall be posted in the break-area trailer.

If chemical contamination exists, work zones will be divided, as suggested in *Occupational Safety and Health Guidance Manual for Hazardous Waste Site Activities*, NIOSH/OSHA/U.S. Coast Guard/USEPA, November 1985, into three zones: EZ, CRZ, and support zone (see Figure 10.3).

For work areas that have both radiological and chemical contamination, the restricted areas will be designated accordingly.

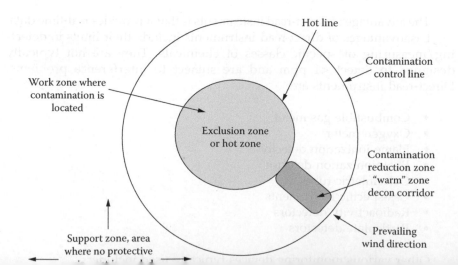

10.8.1 Exclusion zone

The EZ is, in general, the area where chemical, physical, or other hazards occur/exist during project work. All employees are required to follow established procedures, such as wearing the proper PPE, when working in these areas. The location of the EZ should be identified by fencing or other appropriate means. A daily entry log that records the time of entry and exit from the EZ for each person is kept. Unauthorized personnel are not to be allowed in these areas. An EZ may also be identified by radiological postings.

10.8.2 Contamination reduction zone

Personnel and equipment decontamination is performed in the CRZ. All personnel and small equipment entering or leaving the EZ pass through the CRZ to prevent cross-contamination and for the purpose of accountability. PPE is removed in the CRZ, cleaned, and properly stored or disposed of. Drums for handling contaminated trash and reusable PPE are maintained in the CRZ until disposition is determined. Each drum must be labeled as to the appropriate contents of the drum. All water generated from equipment and personal decontamination are also contained on-site and disposed of in an appropriate manner.

At each CRZ, appropriate monitoring equipment shall be available for personnel to frisk themselves for the presence of contamination prior to their leaving the CRZ.

For large equipment, an equivalent CRZ should be designated where the equipment is sampled/monitored for contamination, and decontaminated as required, prior to leaving a contaminated area.

10.8.3 Support zone

The support zone, or clean zone, is the area outside the EZ, CRZ, and within the geographic perimeters of the site. The support zone is used for staging of materials, parking of vehicles, office facilities, sanitation facilities, and receipt of deliveries. Eating, drinking, and smoking are allowed only in designated areas of the support zone.

10.8.4 Emergency entry and exit

During an emergency, personnel evacuate to a predetermined location at the

10.9   Decontamination

Decontamination is performed to stop the spread of contamination and to ensure control of the hazardous waste site.

Physical decontamination: Relies on physical means of removal (brushes, water sprays, freezing, steaming, scraping off of contaminants, etc.).

Chemical decontamination: Takes hazardous chemical to a less hazardous stage; may neutralize chemicals; relies on chemical solutions to decontaminate.

Decontamination of equipment and personnel is also performed to reduce worker risks. Decontamination will generally occur at the edge of an EZ, contamination area, or radiation area. Additional, temporary decontamination stations may be established as project activities and needs warrant. In general, everything that enters a restricted area at the site must either be decontaminated or properly discarded upon exit from the EZ. Everything that leaves a contaminated area at the project must be frisked to determine if contamination is present, and if it is, either be decontaminated or properly discarded. Personnel decontamination consists of discarding disposable PPE, cleaning reusable PPE, and washing the hands and face. If appropriate, personnel shall frisk themselves for radioactive contaminants following decontamination. *Frisking* is the act of monitoring for the presence of contamination by holding an appropriate radiation detection probe 0.25 in. to 0.5 in. from the surface to be frisked, and moving the probe at approximately 2 in./sec. A rise in meter count rate will indicate when contamination is found. All personnel must enter and exit an EZ and radiologically posted areas through a CRZ.

Bibliography

1. American Conference of Governmental Industrial Hygienists (ACGIH), Threshold Limit Values and Biological Exposure Indices, 1999.
2. Nuclear Regulatory Commission, Regulatory Guide 1.86, Termination of Operating Licences for Nuclear Reactors, Washington, D.C., 1974.
3. National Safety Council, *Fundamentals of Industrial Hygiene*, 1996.
4. NIOSH DHHS Publication No. 96-132. The Effects of Workplace Hazards on Male Reproductive Health, Washington, D.C.
5. *Title 10 Code of Federal Regulations (CFR), Part 20*, Standards for Protection against Radiation, Washington, D.C.
6. *Title 10 CFR, Part 19*, Notices, Instructions and Reports to Workers: Inspection and Investigations, Washington, D.C.
7. *Title 29 CFR, Part 1910*, Safety and Health Regulations for General Industry, Washington, D.C.

Chapter 10:   Establishing project hazards and safety control measures   *187*

10. USACE, *Safety and Occupational Health Document Requirements for Hazardous, Toxic and Radioactive Waste (HTRW) and Ordnance and Explosive Waste (OWE) Activities*, Appendix B. ER 285-1-92, Washington, D.C., 1994.
11. U.S. Department of Health and Human Services, Public Health Service, Centers for Disease Control and Prevention, National Institute for Occupational Safety and Health (NIOSH et al.), *NIOSH Pocket Guide to Chemical Hazards, NIOSH Publication No. 97-140, Washington, D.C., June 1997.*
12. U.S. EPA 1988. *Safety Operating Guidelines*, Washington, D.C., July 1988.
13. U.S. EPA 2000. *Multi-Agency Radiation Survey and Site Investigation Manual (MARSSIM)*, NUREG-1575, Rev. 1, Washington, D.C., August 2000.

chapter 11

Cost, productivity, and scheduling issues

11.1 Contracting options

The federal acquisition regulation (FAR) provides a good starting point to understanding contract options in the government or commercial environmental restoration arena. Although the latter does not have any legal requirements to adhere to FAR, the principles and issues relative to environmental restoration projects are consistent and, as such, government contracting philosophies provide a relevant starting point, after which applicable commercial contracting nuances and practices will also be discussed.

There are four general contract types available under FAR: fixed price, indefinite quantity, time and material, and cost reimbursement.

Fixed contractors in turn can be divided into four subtypes: (1) firm-fixed price — FAR 16.202; (2) unit price — FAR 16.2 and 12.403; (3) fixed price incentive — FAR 16.204 and 16.403; and (4) fixed price with award fee.

Under FAR, a firm-fixed price contract may be sealed or negotiated based on defined design or performance specifications. A firm-fixed price contract is not subject to change despite contractor performance experience. Thus, all the financial risk is on the contractor. Given the high degree of uncertainty involved with macroengineering-sized environmental restoration projects and the emphasis on flexible, observational-approach-based decision making, firm-fixed price is not a typical option for a macroengineering type of project.

A unit price contract may also be sealed or negotiated. The required quantity of the "unit" can be undetermined but, typically, a "variation in estimated quantities" clause is available based on reasonably different design or performance specifications to allow appropriate adjustment between estimated quantities and the actual quantities deliverable. This can be a critical point when technologies with well-defined quantity thresholds are under consideration. Also, from a contractor agency/organization standpoint, unit

190   Macroengineering: An environmental restoration management process

Under FAR, fixed price inventive contracts are restricted to a negotiated procurement route. Fixed price incentive is the preferred contract option when cost uncertainties exist, but there is a clear potential for cost reduction or performance gains. Macroengineering projects will typically have strong opportunities for this type of contract vehicle. The enhanced contractor flexibility and responsibility typical in fixed price incentive contracts often times result in positive cost reduction performance gains that are driven by the positive profit incentive of potentially stronger profit margins for the innovative and successful contractor.

Under FAR, fixed price with award fee contracts are awarded by sealed bid only. At the start of the contract, the terms are a firm fixed price for a definitive specification with additional fee or partial fee available for exceptional performance based on objective measurement and criteria established via a clear, unambiguous evaluation process.

Under FAR, the indefinite quantity type contract may be used with a sealed bid or a negotiated procurement. An indefinite quantity contract is set when it is impossible to determine in advance an accurate estimate of the quantities of supplies or services that will be needed to complete designated activities during the contract period (although a quantity estimate may be provided for bid evaluation purposes). To maintain control and protect the awarding organization, an indefinite quantity type of contract should state:

1. The method of ordering work
2. The minimum/maximum orders allowable during a specified period
3. A fixed unit price schedule as a basis of cost for items to be ordered

There are two types of indefinite quantity contracts:

1. A requirements type (per FAR 16.503)
2. An indefinite quantity type (per FAR 16.504)

In a requirements-type contract, the contracting organization is not obligated to place any minimum orders. However, the contract does obligate the awarding organization to order from a successful contractor and not from other sources for all supplies and services described in the contract. The indefinite quantity subtype provides for a stated minimum and maximum amount to be ordered by the awarding organization during the contract period. Under a macroengineering approach, these are typically the type of vehicles used for specialty second tier contracted support.

Under FAR, time and materials contracts may be negotiated

unit cost. The contract may contain estimated quantities to be used for evaluation purposes. However, the awarding organization should be aware of potential dangers if ceilings are not established, particularly for materials handling if there is a potential for material expansion due to technology options. Time and cost standards, applicable to particular work items relevant to the contract should be established and actively monitored by the awarding organization.

Under FAR, cost reimbursement contracts are used only for negotiated procurement. The total award fee plus base fee is limited by statutory limits [FAR 15.903(d)]. This is the most costly contract type for an awarding agency to administer and requires the contractor organization to demonstrate a strong accounting system. This contract is typically only used when the nature of the work or cost estimate unreliability makes it unfeasible to use any other contract type. There are two such types of cost reimbursement contracts. They are: (1) cost plus incentive fee (FAR 16.404-1) and (2) cost plus award fee (FAR 16.402.2). Cost plus incentive fee contracts are very suitable for research and development project types, particularly when development has a high probability. The cost plus incentive fee contracts must contain the target cost, target fee, as well as minimum and maximum fees, and a fee-adjustment formula. The performance incentives must be clearly established and able to be objectively measured. Fee adjustment is made at the project completion based on end results. Cost plus award fee contracts are utilized when it is impossible to write a precise description of the work expected to be performed. It is used when a contract completion is feasible, but although incentives are desired, it is not possible to measure definitely. The incentive fee evaluation typically is subjective and not subject to dispute.

Table 11.1 provides an example of the type of structures in which specific contract types may be appropriate.

Table 11.1 Recommended Acquisition Strategies for Hazardous Waste Remediation

Remediation	Fixed price	Indefinite quantity	Time and materials	Cost reimbursement
Simple earth-moving restoration	√			
Complex environment restoration	√	√	√	
Simple pump and treat	√			
Complex pump and treat	√	√	√	√
Simple soils and sludge	√			

11.2   *Developing cost and schedule estimates*

Environmental remediation cost estimating, unlike standard construction and industrial engineering cost estimation, is still in a stage of development. The validity of environmental remediation cost estimates may be subject to debate because of the long-term nature of cleanup processes, the unproven remediation technologies used, and the uncertainty regarding ultimate cleanup levels. Further complications result from inconsistent presentation, interpretation, and application of the available environmental remediation cost database. Therefore, the estimates are likely to include a large margin of error. As a result, there is a tendency to subdivide project activity into what appears to be more a manageable size. However, these arbitrary sub-divisions may result in lost opportunities for economies-of-scale cost savings and inefficiencies in approach. In particular, as will be shown in this section, taking a macroengineering approach can possibly diminish the cost impact criticality of cleanup-level decisions. Also, it is critical to quickly interface projected system design resource needs within the "limitations of existing programmatic funding." This allows for optimizing the implementability of proposed schedules by clearly identifying those issues that need special funding and identifying where budget constraints dictate scheduling con-straints.

The following presents common assumptions, influences on, and breakdown of costs for macroengineering-type remediation. Costs can be separated into four categories: capital expenditures, labor, allowances, and analytical costs. Costs should be developed assuming reasonably achiev-able industrial operating standards. Typically, given the developing state of environmental remediation technology, initial macroengineering cost estimates do not take into account factors such as efficiencies gained in processing as cleanup proceeds. There may be significant potential savings in this regard.

For example, the cost estimates presented herein take no advantage of possible cost efficiency adjustments to worker safety requirements (presently assumed to require a high level of protection) and no significant advantage of the utilization of robotics. Savannah River's mixed-waste management facility represents a case study of these types of savings. The initial cost estimates for RCRA closure using dynamic compaction and capping of Savannah River's mixed-waste management facility totaled $118 million (1989 dollars). This included design, procurement, and remediation. The estimate was subsequently revised downward to $52.8 million when it was deemed that a higher level of worker protection was not necessary. It was further revised downward to less than $18 million due to a decrease in the

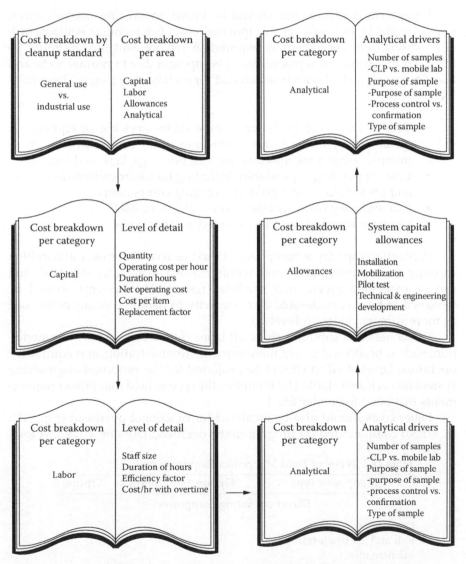

Macroengineering costs should provide a micro level of detail for the "entire" effort.

Figure 11.1 Typical cost model components.

They should include costs of equipment (i.e., bulldozers, trucks

Equipment cost estimates should be based on existing market prices, where available, or based on extrapolated costs from similar equipment if new (specialized) equipment is required. A replacement factor should be included to account for replacement of equipment due to normal wear and tear. Operating costs should be estimated for each hour of usage. These costs include:

- Cost of labor and parts for routine daily servicing of equipment, including repairing and/or replacing small components such as pumps, carburetors, injectors, motors, filters, gaskets, and worn lines
- Cost of operating expendables including fuel, lubrications (filters, oil, and grease), tires, and ground-engaging components
- Extraordinary costs specific to the equipment itself
- Operator wages should be included in labor costs

Production capacity assumptions should be based on material handling capacity assumptions that are well within reasonable industry standards and can be expanded given their modular nature. The systems should be designed to have considerable extra capacity to handle contingencies such as more stringent cleanup levels.

Personnel costs should include all labor costs associated with remediation, such as health and safety, management/administration, and equipment operators. Level of effort should be projected for the proposed engineering system for each area. Table 11.2 identifies the type of field manpower requirements that need to be identified.

Labor costs should also be escalated for functional overhead (e.g., 30% or × 1.30) compounded by programmatic overhead (30% or × 1.30). A total

Table 11.2 Types of Field Manpower Requirements

Manpower type	Option 1	Option 2	Option 3
Direct operating manpower			
Operations			
Rail and/or truck transport			
Maintenance			
Support personnel			
Administration			
Quality assurance			
Decontamination			
Health physics technician			

Chapter 11:   Cost, productivity, and scheduling issues   *195*

of 2080 h per worker per year is assumed. Most operations assume operating two shifts during the summer or 1.5 shifts for the entire year. Typically, CERCLA remediation sites achieve a shift productivity of 6 h per 8 h; however, a labor productivity assumption of 5 h productive labor per 8-h shift is a valid conservative industry assumption. However, beware that this may require review per renegotiation of existing labor work rules at specific sites (particularly true at DOE sites). Utilizing a 5 h per 8-h productivity assumption provides for start up, transportation, decontamination, work break, and lunch issues, as well as downtime due to inclement weather and equipment breakdown.

Schedules should be developed in a manner that allows acceleration based on the requirements of the company and resources utilized. Basic issues to be considered are identified in Figure 11.2.

The macroengineering systems developed should have reserve throughput capacities that allow their system components to meet unforeseen expansion of expected waste volume without impacting overall schedule. Furthermore, the proposed systems should be modular in nature if only to allow additional

	2000	2001	2002	2003	2004	2005	2006	2007	2008	2009	2010
OPERABLE UNIT AREA											
NEPA-EIS Impact Study	■										
Design, Permit, Program Development	■	■									
Engineering Developing and Testing		■	■								
Procurement Activities			■	■							
Construction & Field Mobilization				■	■						
Site Excavation/ Demolition					■	■					
Engineering Development & Testing						■	■				
Operational								■			

flexibility for dealing with unexpected events and for enhancing the remediation pace where project and regulatory controls permit. Thus, the macroengineering approach should offer a strong baseline schedule that can be
continually measured for progress and adapted to meet contingencies.

In contrast, the microengineering approach for source area remediation
is subject to a wide variety of delays and false starts. Furthermore, the
microengineering approach, for all practical purposes, cannot provide a
baseline to measure progress on sitewide remediation as a whole. Given the
current system's penchant for a multiple study/review process, a high potential exists for delay from a regulatory and budgeting standpoint. Currently,
each record of decision (ROD) is projected to take 3 to 7 yr to complete.

It is recognized that the macroengineering approach is also subject to
potential schedule problems. For federal sites, the NEPA process is a critical
path to implementation. However, the NEPA issues can be mitigated if the
schedule calls for NEPA documentation to be done in parallel with design
and preconstruction activities. Typically, NEPA documentation is completed
prior to commencement of design and preconstruction activities. However,
this approach typically results in at least a 5-to-10-yr delay to the overall
schedule. The financial risk of proceeding with up-front design and preconstruction activities is oftentimes outweighed by the favorable schedule
impact of paralleling these activities, not to mention the long-term costs of
extending remediation past activities well into the 21st century. Also, RCRA
permitting documents have been accepted in current cases in lieu of NEPA
documentation and resulting in exception of NEPA documentation as unnecessary and duplicative. Also, many organizations proceed with CERCLA
documents in a joint fashion with their NEPA documentation.

11.3   *Productivity, cost, and schedule issues*

The productivity, cost, and schedule estimates are based on assumptions of
certain operations being performed and are affected by certain cost and schedule drivers. Changes in these elements will cause projected costs and schedules
to change. Following is a discussion of major cost components and cost drivers
in each area in macroengineering. Productivity, cost, and schedule projections
are based on a number of parallel operations occurring simultaneously:

- Overburden removal operations
- Excavation/demolition operations
- Sorting operations
- Mobile laboratory operation

- Disposal operations
- Restoration and closure operations

These operations influence each other, and any given bottleneck can disrupt the overall productivity. Other major productivity cost and schedule drivers for unique remediation problems include:

- The availability of shipping containers for specialty wastes, such as high-activity-level radioactive wastes
- Specialized containment system costs
- Underwater pipeline excavation and removal if offshore sediments are found to be contaminated
- Buried-waste excavation if significant quantities of intact drums are found
- Buried-waste excavation where wastes encountered present highly explosive or flammable hazards
- Substantial differences in waste forms encountered from that expected
- Serious problems uncovered during treatability studies and proto-type equipment development or demonstrations

For example, the utilization of single-use shipping containers for unique waste streams, which would primarily be driven by waste-handling/retrievability requirements, is typically the largest (and most volatile) cost driver. This reflects the fact that often the key driver for container cost is the cost for certification of the container type. For sites with unique container storage requirements, this certification is frequently done for a limited number of containers of a particular container design. However, given the scale of the typical macroengineering site, there may be an opportunity to generate sufficient standardization of containers to alleviate this up-front cost if large-scale, sitewide remediation planning is conducted.

If temporary containment structures are required (e.g., bubbles), they too will be prime cost drivers. Containment systems are typically expensive to build, operate, and maintain. However, the sheer size of the structures makes for expensive construction, more so in material costs than in labor. In addition, large containment structures would require high-capacity ventilation systems. Although the ventilation system would consist of conventional components, they could be expensive to build as the size of the structure goes up. In addition, ventilation systems will also consume large quantities of HEPA filters, a continuing operating cost.

Underwater pipeline removal costs are also potential cost drivers. They

As another example, encountering large numbers of intact drums would potentially slow down the buried-waste excavation operations for the surficial contaminant soil, mining technology approach. The latter is more suited for surficial containment soil structures. Even though intact drums could subsequently be handled off-line from the excavation under the mining approach, the unearthing of drums would have to be done slowly and carefully to preserve the integrity of intact drums. Rather than using large-capacity loaders for excavation, small-scale, one-by-one drum handlers may have to be used. Although this is technically achievable with the proposed system, costs would increase as a result of slower excavation rates.

Materials such as pressurized drums, drums containing hydrogen (from radiolysis), highly flammable organics, compressed gas cylinders, and unexploded ordnance/munitions could also pose additional requirements for special handling procedures that may slow excavation.

Productivity can also be impacted by management decision-making issues, including:

- Lengthy delays in RCRA/CERCLA or NEPA processes
- Budget restrictions
- Delays in the contractor-selection process

The most significant cost driver in groundwater remediation will be the selection of options. There is an inherent tradeoff on groundwater remediation options. On one hand, there is the very understandable desire to immediately attack the problem. On the other hand, there are considerable advantages to adopting a defensive maintenance posture and taking maximum advantage of natural remediation processes and natural radiation decay processes (if radiation is an issue). As shown in Figure 11.3, groundwater options are subject to a wide fluctuation dependent on scenario-cleanup objectives and point of compliance.

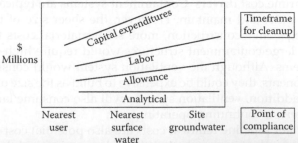

Dependent upon timeframe for cleanup, the point of compliance, and the type of environmental program

Chapter 11:   Cost, productivity, and scheduling issues   *199*

Table 11.3 Aggregate Area Macroengineering Productivity Requirements
(General Use)

	Productivity rate	Operations
Mining approach	350 yd³/h	Remove, sort, package, and transport
Contained hazardous waste management approach	500 yd³/h	Accept, sort, package, final dispose, and barrier
Industrial approach	100 yd³/h	Remove, treat, sort, package, and transport

The more stringent the time frame objectives for cleanup and the tighter the three-dimensional geographic point for compliance, the more dramatic the rise in costs. Typically, the latter are very capital intensive, requiring considerable outlays in treatment facility costs and specialty piping costs. End points have also been added that show what costs may be involved if only a passive monitoring program is maintained vs. an aggressive cleanup, such as a sitewide groundwater-mining scenario, is attempted. The former is based on assuming a limited monitoring program.

Lastly, it is recognized that sound management practices may dictate a phased approach to macroengineering. The proposed system has the flexibility to accept such a phased approach; however, it must be accomplished in a manner consistent with the overall macroengineering philosophy, taking advantage of economies of scale and minimizing redundancy.

Table 11.3 presents rule-of-thumb production rates for macroengineering systems under the general-use cleanup option for the various scenarios previously.

Under this option, the mining technology approach, the system must remove, sort, package, and transport waste material at a rate of over 300 yd³/h. This rate compares favorably with surface mining and major civil engineering operations that can handle thousands of cubic yards per hour.

The industrial approach includes an assumption of an 80% waste volume reduction step as well as removal, sorting, packaging, and transporting. In this case, the system is expected to process material at a rate of 100 yd³/h. Again, this is a highly manageable production figure for excavation and treatment facility requirements. Given the modular nature of systems such as soil washing, the anticipated production rate can be met and can be expanded.

Under the general-use scenario, the hazardous waste management system approach calls for disposal of hazardous waste material at a rate of over 500 yd³/h. This is well within the established range of production capabilities for disposal facilities.

Table 11.4 System Operational Cost and Capital Allowances

System operational cost allowances		
Allowance category	Percentage[a]	Description
Engineering design support	10%	All contractor costs associated with engineering design support prior to and during construction and operations
Supplies and materials	1%	All general supplies and materials including office and construction (small tools, safety clothing, hygiene equipment, etc.)
Utilities	15%	Electricity, water, and fuels at each construction site including associated labor costs
Maintenance	15%	Allocation of downtime and major off-site maintenance required on equipment and facilities maintenance
Waste disposal from operations	10%	Cost of off-site disposal including pickup, processing, and disposal
Decontamination and decommissioning	10%	Cost of decontamination of equipment, personnel, capital, and labor
Permitting	10%	Permitting costs and factor to account for possible delays in permitting process
Contingency	25%	Factor for unforeseen problems that could lead to additional capital costs or delays in progress
Total	96%	
System capital expenditure allowances		
	Percentage[b]	Description
Installation costs	10–25%	All costs associated with construction of major equipment and

Table 11.4 System Operational Cost and Capital Allowances (Continued)

System capital expenditure allowances		
	Percentage [b]	Description
Pilot test costs	10–25%	All costs for testing system throughout assumption and system components
Technology and engineering development costs	10–25%	All costs for enhancing existing technologies
Total	40–100%	

[a] Percentage of capital costs and personnel costs.

[b] Percentage of capital costs.

remediation contractors. The allowances also reflect the high degree of flexibility and modification that may be required for the remediation systems.

The allowances should be broken down into two major categories, which in turn should be broken down into more specific subcategories. An example of system operational cost allowances, based on a percentage of capital expenditures and personnel costs, is presented in Table 11.4. An example of system capital expenditure allowances, based on a percentage of capital expenditures, is also presented in Table 11.4.

Table 11.5 presents EPA's recommended fee allowances for various types of facility closures. These base allowances assume minimal environmental

Table 11.5 Typical Fees for Closure and Postclosure Recommended by the EPA

Facility type	Engineering fee (%)	Contractor's overhead and profit (%)	Contingency fee (%)		Total [b]
			Closure	Postclosure	
Container storage	10	25	10	—	45
Treatment or storage tank	10	25	15	—	50
Incineration	15	25	15	—	55
Waste pile	10	25	15	10 [a]	60
Surface impoundment	15	25	25	15 [a]	80
Land treatment	10	25	20	10	65
Landfill	15	25	25	15	80
Multiple	15	25	25	15 [a]	80

202   *Macroengineering: An environmental restoration management process*

degradation and vary from 45% to 80%, depending on the type of facility. The EPA caveats these recommendations by warning that contingency fees and allowances can be as high as 200% for sites with known soil contamination, obviously dependant on the degree and impact of the contamination.

Given the degree of contamination addressed in this study, the highest levels of contingencies and allowances were factored into development of the cost estimate.

Bibliography

1. U.S. Congress, Office of Technology Assessment. Complex Cleanup: The Environmental Legacy of Nuclear Weapons Production, OTA-0-484, Washington, D.C., February 1991.
2. William, R. and Zobel, P.E., Acquisition Selection for Hazardous Waste Remediation, U.S. EPA, Hazardous Site Control Division, Design and Construction Issues at Hazardous Waste Sites Conference Proceedings, May 1–3, 1991, EPA/540/8-91/012, Washington, D.C.

chapter 12

Summary

In summary, the possible advantages of a large-scale aggregate (macroengineering) approach over the traditional RFI/CMS and RI/FS process for the cleanup of large-scale environmental restoration areas are primarily reduced cost and time of operations. These reductions in time and cost may be achieved through the use of large-scale construction and mining equipment guided by rapid, real-time characterization and monitoring procedures. It should be noted that using the large-scale remediation approach, the impact of cleanup level to cost and schedule is mitigated for the soil and solid hazardous material remediation activities. This, along with the reduced need for manpower and a faster cleanup, could result in an overall decrease in exposure for humans and the environment.

In addition, macroengineering provides a conceptual, baseline-remediation system to:

- More accurately assess the economic impact and resource requirements of remediation activities for large-scale site restoration activities
- Serve as quicker, more cost effective, and technically more viable
- Act as useful benchmark to evaluate the feasibility of future technology alternatives and serves as a foundation upon which to base further planning

The potential disadvantages of the macroengineering approach include:

- High up-front capital investment required for remediation equipment
- The design and cost of disposable and reusable containers
- Fugitive dust control and the potential spread of contamination during the treatment, packaging, and transfer of contaminated materials, requiring significant engineering design work

The capital investment makes it imperative that the approach be

204   *Macroengineering: An environmental restoration management process*

is need for pilot-testing specific technical approaches; however, piecemeal implementation must be avoided to achieve the full benefits of economies of scale inherent in this concept.

The intent of the baseline study should not be to establish a final design but to evaluate a variety of approaches that will encourage a wide range of alternative solutions. The baseline engineering system serves as a useful tool for decision makers and the general public for future planning.

Index

A

AA/AE, *see* Atomic absorption/emission spectrophotometry (AA/EE)

accuracy, 70, *see also* Precision

acid washing, industrial-engineering approach, 33

action levels (ALs), 133–135

additive effects, 138, 140

advanced geophysical data processing, 77–78

agreement stage, 151–152

air drill systems, 62, 66

airlock entrances, 26, 35

ALARA, *see* As Low As Reasonably Achievable (ALARA)

alert suppression systems, 19

allowances, 192

alpha particles, 175–177

American Society of Testing and Materials (ASTM), 62, 114

analysis process, mobile laboratories, 87, *88*, 89–96

analytical capabilities, mobile laboratories, 85–87

analytical costs, 192

Analytical Products Group (APG), 99

ancillary support systems, 26, 35

animal species, 147–148, 159

anionic species, 25

ANSI Z88.2, 179

APG, *see* Analytical Products Group (APG)

Applicable or Relevant and Appropriate Requirements (ARARs)

  natural attenuation remedy, 131

ARARs, *see* Applicable or Relevant and Appropriate Requirements (ARARs)

ARCHIE model, 167

area-sampling equipment, 25

Army Corps of Engineers, *see* U.S. Army Corps of Engineers (COE)

aromatics screening, *97*

ash pits, 34

As Low As Reasonably Achievable (ALARA), 27, 177

ASTM, *see* American Society of Testing and Materials (ASTM)

atomic absorption/emission spectrophotometry (AA/EE), 93

attrition scrubbing, 33

auxiliary equipment, mobile laboratories, 95, *96*

B

backfilling operations, 27, 29, 33

background values, 74

backhoes, 31

bare-bones cleanup treatment and disposal scenarios, 112

barrier systems, 38, *38*

baseline engineering document, *see* Preconceptual engineering baseline study

BDAT, *see* Best-demonstrated available technologies (BDAT)

bench investigation, 55–56

best acceptable technology (BAT), 13

best-demonstrated available technologies (BDAT), 124

Index 207

Index   *209*

Index 211

Index 213

Index 215

Index 217

T - #0175 - 071024 - C0 - 234/156/12 - PB - 9780367453688 - Gloss Lamination